BIOMECHANICS OF
HUMAN MOVEMENT

BIOMEDICAL ENGINEERING AND HEALTH SYSTEMS:

A WILEY-INTERSCIENCE SERIES

Advisory Editor: JOHN H. MILSUM, University of British Columbia

BIOMECHANICS OF HUMAN MOVEMENT

DAVID A. WINTER

University of Waterloo
Waterloo, Ontario, Canada

A WILEY-INTERSCIENCE PUBLICATION

JOHN WILEY & SONS
New York • Chichester • Brisbane • Toronto

Library of Congress Cataloging in Publication Data

Winter, David A   1930–
    Biomechanics of human movement.

    (Biomedical engineering and health systems)
    "A Wiley-Interscience publication."
    Includes bibliographical references and index.

1. Human mechanics. 2. Kinesiology. I. Title.

QP303.W59   612'.76   79-12660
ISBN 0-471-03476-2

Printed in the United States of America

10 9 8 7 6 5 4 3 2 1

To my family, colleagues, and students

all of whom have encouraged and
influenced me over the years.

Series Preface

The provision of good universal health care has only just become a social imperative. Like such other interlocking systems as transportation, water resources, and metropolises, the recognition of the need to adopt a systems approach has been forced upon us by the evident peril of mutual ruin resulting if we do not. The required systems approach is necessarily interdisciplinary rather than merely multidisciplinary, but this always makes evident the very real gulf between disciplines. Indeed, bridging the gulf requires much more time and patience than necessary just to learn the other discipline's "language." These different languages simultaneously represent and hide the whole gestalt associated with any discipline or profession. Nevertheless sufficient bridging must be achieved so that interdisciplinary teams can tackle our complex systems problems, for such teams constitute the only form of "intelligence amplification" that we can presently conceive.

Fortunately the existence of urgent problems always seems to provide the necessary impetus to work together in the same direction, using St. Exupery's thought, and, indeed, as a result of this to understand and value each other more deeply.

This series started with its emphasis on biomedical engineering. It illustrates the application of engineering in the service of medicine and biology. Books are required both to educate persons from one discipline in what they need to know of others, and to catalyze a synthesis in the core subject generally known as biomedical engineering. Thus one set of titles will aim at introducing the biologist and medical scientist to the quantitatively based analytical theories and techniques of the engineer and physical scientist. This set covers instrumentation, mathematical modelling, signal and system analysis, communication and control theory, and computer simulation techniques. A second set for engineers and physical scientists will cover basic material on biological and medical systems with as quantitative and compact a presentation as is possible. The omissions and simplifications necessitated by this approach should be justified by the increased ease of transferring the information, and more subtly by the increased pressure this can bring to bear upon the search for unifying quantitative principles.

A second emphasis has become appropriate as society increasingly demands universal health care, because the emphasis of medicine is now rapidly shifting from individual practice to health care systems. The huge financial burden of our health care systems (currently some 8% of GNP and increasing at about 12% per year) alone will ensure that engineers will be called upon, for the technological content is already large and still increasing. Engineers will need to apply their full spectrum of methodologies and techniques, and, moreover, to work in close collaboration with management professionals, as well as with many different health professionals. This series, then, will develop a group of books for mutually educating these various professionals so that they may better achieve their common task.

Vancouver, British Columbia   JOHN M. MILSUM

Preface

This text has been written to fill a gap in the subject area of biomechanics of human movement. The emphasis is on the assessment of dynamic movements and on realistic movements using "live" data. A wide spectrum of measurement and analysis techniques are described and are aimed at readers interested in higher level quantitative assessments. The text is intended to appeal to the practitioner as well as the researcher and to those concerned with the elite athlete as well as the physically handicapped.

A major characteristic of this text is the format for analyzed examples and student problems. Live data have been used in Chapters 2 to 5 to illustrate and reinforce each stage of the biomechanical analyses. The data provided have come from human gait studies, and the chapters are so arranged that the answers to problems in one chapter form the input to problems in later chapters. Computer listings are provided so that the answers can be verified at each stage and an initial mistake is not carried through an entire series of calculations.

In the last two chapters, on muscle mechanics and electromyography, no student problems are given. Here the aim has been to report as concisely as possible the state of knowledge and its biophysical basis. In no chapter is an exhaustive bibliography given, in most cases only the more significant or representative papers being reported. It is realized that a comprehensive reading list could fill an entire book, and apologies are given to those whose work I have not reported. The author's views on the state of the art will become evident, and areas that are controversial or in error are duly noted and in some cases discussed in detail.

It is expected that the student has had a basic course in mechanics and anatomy, and it would be useful, though not mandatory, if he has also taken calculus and basic electrical science. The major programs to which the book is directed are: physical education, kinesiology, physical and occupational therapy, and bioengineering (rehabilitation engineering).

The preparation of the manuscript for this text has not been a solo effort. I would like to acknowledge the considerable help I have had from John Cairns and John Pezzack in collecting resource material, such as computer listings, graphical plots, and pen recordings. The patience and accuracy of Gayle Shellard in typing, proofreading, and editing most of the manuscript are gratefully acknowledged. The helpful suggestions of several students and faculty members are much appreciated.

DAVID A. WINTER

Waterloo, Ontario, Canada
September 1979

Contents

5 MECHANICAL WORK, ENERGY AND POWER 84

Appendixes

A ANTHROPOMETRIC, KINEMATIC AND FORCE PLATE DATA

B KINETIC ANALYSES

BIOMECHANICS OF
HUMAN MOVEMENT

CHAPTER ONE
Biomechanics as an Interdiscipline

1.0 INTRODUCTION

Biomechanics of human movement can be defined as the interdiscipline which describes, analyzes, and assesses human movement. A wide variety of physical movements are involved—everything from the gait of the physically handicapped to the performance of a superior athlete. The physical and biological principles that apply are the same in all cases. What changes from case to case is the specific movement task and the level of detail that is asked about the performance of that movement.

The list of professionals and semi-professionals interested in applied aspects of human movement is quite long: orthopedic surgeons, athletic coaches, rehabilitation engineers, therapists, kinesiologists, prosthetists, physiatrists, orthotists, sports equipment designers, and so on. At the basic level, the name given to the science dedicated to the broad area of human movement is kinesiology. It is an emerging discipline blending aspects of psychology, motor learning, and exercise physiology as well as biomechanics. Biomechanics, as an outgrowth of both the life and physical sciences, is built on the basic body of knowledge of physics, chemistry, mathematics, physiology, and anatomy. It is amazing to note that the first real "biomechanicians" date back to Leonardo DaVinci, Galileo, Langrange, Bernoulli, Euler, and Young. All these scientists had primary interests in the application of mechanics to biological problems.

1.1 MEASUREMENT VERSUS DESCRIPTION VERSUS ANALYSIS VERSUS ASSESSMENT

The scientific approach as applied to biomechanics has been characterized by a fair amount of confusion. Some descriptions of human movement have been passed off as assessments, some studies involving only measurements have been falsely advertised as analyses, and so on. It is therefore important to clarify these terms. Any quantitative assessment of human movement must be preceded by a measurement and description phase, and if more meaningful diagnostics are needed, a biomechanical

1

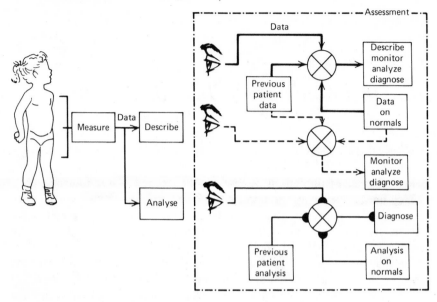

Figure 1.1 Schematic diagram showing the three levels of assessment of human movement.

analysis is usually necessary. Most of the material in this text is aimed at measurement, description, and analysis, supplemented by examples of assessment.

Figure 1.1, which has been prepared for the assessment of the physically handicapped, depicts the relationships between these various modes of assessment. All levels of assessment involve a human being, based on his visual observation of a patient or subject, his data, or some resulting biomechanical analysis. The primary assessment level uses direct observation, which places tremendous "overload" even on the most experienced observer. All measures are subjective and almost impossible to compare with those obtained previously. The observer is then faced with the tasks of documenting (describing) what he sees, monitoring changes, analyzing the information, and diagnosing the causes. If measurements can be made during the patient's movement, then data can be presented in a convenient manner to quantitatively describe the movement. Here the assessor's task is considerably simplified. He can now quantitatively monitor changes, carry out simple analyses, and try to reach a more objective diagnosis. At the highest level of assessment the observer can view biomechanical analyses that are extremely powerful in diagnosing the exact cause of the problem, compare these analyses with the normal population, and monitor their detailed changes with time.

The measurement and analysis of an athletic event can be quite similar to that of an amputee's gait. However, assessing the optimization of energy of the athlete is quite different from assessing the stability of the amputee. The athlete is looking for very detailed and minor changes that will improve his performance by a few percentage points, sufficient to move him from fourth to first place. His training and exercise programs and reassessment normally continue over an extended period of time. The amputee, on the other hand, is looking for improved and safe performance, but not fine differences. He is quite happy to be able to walk at 90 percent of his maximum capability, although the techniques are available to permit him to train and have his prosthesis readjusted until he reaches maximum.

1.1.1   Measurement and Description

It is difficult to separate the two functions of measurement and description. However, for clarity the student should be aware that a given measurement device can have its data presented in a number of different ways. Conversely, a given description could have come from several different measurement devices.

Earlier biomechanical studies had the sole purpose of describing a given movement, and any assessments that were made resulted from visual inspection of the data. The description of the data can take many forms: pen recorder curves, plots of body coordinates, stick diagrams, statistical measures of temporal, or cadence variables. A movie camera, by itself, is a measurement device, and the resulting plots form the description of the event in time and space. In Figure 1.2 we see a system incorporating a cine camera and two different descriptive plots. The coordinates of key anatomical landmarks can be extracted and plotted at regular intervals in time. Time history plots of one or more coordinates are useful in describing detailed changes of a particular landmark and also can reveal to the trained eye changes in velocity and acceleration. A total description in the plane of the movement is the stick diagram in which each body segment is represented by a straight line or stick. Joining the sticks together gives the spatial orientation of all segments at any point in time. Repetition of this plot at equal intervals of time gives a pictorial and anatomical description of the dynamics of the movement. Here, trajectories, velocities, and accelerations can be visualized. To get some idea of the volume of the data present in a stick diagram the student should note that one full page of coordinate data (Table A.4) is required to make this complete plot for the description of the event. The coordinate data can be used directly for any desired analysis: reaction forces, muscle moments, energy changes, efficiency, and so on. Conversely, an assessment can sometimes be made directly from the description. A trained observer, for example, can scan a

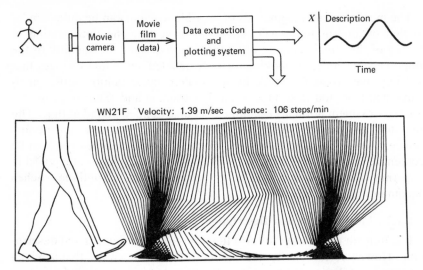

WN21F   Velocity: 1.39 m/sec   Cadence: 106 steps/min

Figure 1.2   Flow of data from a movie camera and the plotting of these data in two different forms, each yielding a different description of the same event.

stick diagram and extract useful information that will give some directions for training or therapy, or give the researcher some insight into basic mechanisms of movement.

1.1.2   Analysis

The measurement system yields data that are suitable for analysis. This means that data have been calibrated and are as free as possible from noise and artifacts. Analysis can be defined as any mathematical operation that is performed on a set of data to present them in another form or to combine the data with other data to produce a variable that is not directly measurable. From the analyzed data information may be extracted to assist in the assessment stage. In some cases the mathematical operation can be very simple, such as the processing of an electromyographic signal to yield an "envelope" signal (Figure 1.3). The mathematical

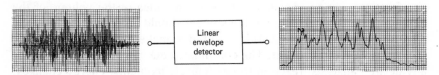

Figure 1.3   An example of a mathematical analysis of an electromyographic signal.

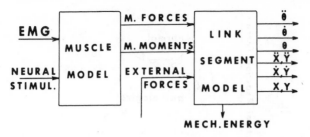

Figure 1.4   Schematic diagram to show the relationship between the neural, kinetic, and kinematic variables required to describe and analyze human movement.

operation performed here can be described in two stages. The first is a full wave detector (electronic term for a circuit that gives the absolute value). The second stage is a low pass filter (which mathematically is a weighting function that follows the average trend of the signal amplitude). A more complex biomechanical analysis could involve a link segment model, and with appropriate kinematic, anthropometric, and kinetic input data we can carry out analyses which could yield a multitude of significant time-course curves. Figure 1.4 depicts the relationships between these variables. The output of the movement is what we see and can be described by a large number of kinematic variables: displacements, joint angles, velocities, and accelerations. If we have an accurate model of the human body in terms of anthropometric variables we can develop a reliable link segment model. With this model and accurate kinematic data we can predict the net forces and muscle moments that caused the movement we have observed. Such an analysis technique is called "input discovery" and is extremely valuable, allowing us to get at variables that cannot be measured directly. In a similar manner the muscle forces and moments are the output of the muscle system. With suitable data concerning the neural input to the muscle we can investigate the muscle system itself and try to develop a reliable mathematical model of the muscle.

1.1.3   Assessment

The entire purpose of any assessment is to make a positive decision about a physical movement. An athletic coach might ask, "Is the me-chanical energy of the movement better or worse than before the new training program was instigated, and why?" Or the orthopedic surgeon may wish to see the improvement in the knee muscle moments of a patient a month after surgery. If the analysis does not cause a decision to be made

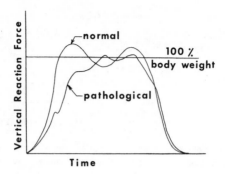

Figure 1.5 An example of a force plate curve that has sometimes been used in the assessment of normal and pathological movement.

it can be said that there is no information present in the analysis. The decision may be positive in that it may confirm that the coaching, surgery, or therapy has been correct and should continue exactly as before. Or, if this is an initial assessment, the decision may be to proceed with a definite plan based on new information from the analysis. The information can also cause a negative decision, for example, to cancel a planned surgical procedure and to prescribe therapy instead.

Some biomechanical assessments have involved a look at the description itself rather than some analyzed version of it. A common example is to look at the ground reaction force curves from a force plate, an electromechanical device that gives an electrical signal which is proportional to the weight (force) of the body acting downward on it. Such patterns appear in Figure 1.5. A trained observer can detect pattern changes as a result of pathological gait and may come to some conclusions as to whether the patient is improving and possibly some reasons why. However, at best, this approach is crude and yields little information regarding the underlying cause of the observed patterns.

1.2   SCOPE OF THE TEXTBOOK

The best way to outline the scope of any scientific text is to describe the topics covered. In this text the biomechanics of human movement has been defined as the mechanics and biophysics of the musculoskeletal system as it pertains to the performance of any movement skill. The neural system is also involved, but it is limited to electromyography and its relationship to the mechanics of the muscle. The variables that are used in the description and analysis of any movement can be categorized as follows: kinematics, kinetics, anthropometry, muscle mechanics, and electromyography. A summary of these variables and how they interrelate now follows.

1.2.1   Kinematics

Kinematic variables are involved in the description of the movement, independent of the forces that cause the movement. They include linear and angular displacements, velocities, and accelerations. The displacement data are taken from any anatomical landmark: center of gravity of body segments, centers of rotation of joints, extremes of limb segments, or key anatomical prominances. The spatial reference system can be either relative or absolute. The former requires that all anatomical coordinates be reported relative to a fixed anatomical landmark, for example, body center of gravity. An absolute system means that the coordinates are referred to an external spatial reference system. The same applies to angular data. Relative angles mean joint angles; absolute angles refer to an external spatial reference, for example, in a two-dimensional system, horizontal to the right is 0°, and counterclockwise is positive.

1.2.2   Kinetics

The general term given to the forces that cause the movement is kinetics. Both internal and external forces are included. Internal forces come from muscle activity, ligaments, or from friction in the muscles and joints. External forces come from the ground (ground reaction forces), from active bodies (those forces exerted by a tackler in football) or by passive bodies (as by wind resistance or by a baseball in the process of being thrown or caught). A wide variety of kinetic analyses can be done. The moments of force produced by muscles crossing a joint, the mechanical power flowing from those same muscles, and the energy changes that result from this power flow are all considered part of kinetics. It is here that a major focus of the book is made, because it is in the kinetics we can really get at the cause of the movement and therefore get some insight into the mechanisms involved. A large part of the future of biomechanics lies in kinetic analyses because the information present permits us to carry very definitive assessments.

1.2.3   Anthropometry

Many of the earlier anatomical studies involving body and limb measurements were not considered to be of interest to biomechanics. However, it is impossible to evolve a biomechanical model without some data regarding masses of limb segments, location of mass centers, segment lengths, centers of rotation, angles of pull of muscles, mass and cross-sectional area of muscles, moments of inertia, and so on. The accuracy of any analysis depends as much on the anthropometric measures as on the kinematics and kinetics.

1.2.4   Muscle and Joint Biomechanics

One body of knowledge that is not included in any of the above categories is the mechanical characteristics of the muscle itself. How does its tension vary with length and with velocity? What are the passive characteristics of the muscle—elastic and viscous? What are the various characteristics of the joints? What are the advantages of double joint muscles? What are the differences in muscle activity during lengthening versus shortening? What kind of mathematical models best fit a muscle? How can we calculate the center of rotation of a joint? The final assessment of the many movements cannot ignore the influence of active and passive characteristics of the muscle, nor can it disregard the passive role of the articulating surfaces in stabilizing joints and limiting ranges of movement.

1.2.5   Electromyography

The neural control of movement cannot be separated from the movement itself, and in the electromyogram we have information regarding the final control signal of each muscle. The electromyogram (EMG) is the primary signal to describe the input to the muscular system. It gives information regarding which muscle activity is responsible for a muscle moment or whether antagonistic activity is taking place. Because of the relationship between a muscle's EMG and its tension, a number of biomechanical models have evolved. The EMG also has information regarding the recruitment of different types of muscle fibers and the fatigue state of the muscle.

CHAPTER TWO
Kinematics

2.0 HISTORICAL DEVELOPMENT AND COMPLEXITY OF PROBLEM

Man's interest in the actual patterns of movement of man and animals goes back to prehistoric times, and was depicted in cave drawings, statues, and paintings. Such replications were subjective impressions of the artist. It was not until a century ago that the first motion picture cameras recorded locomotion patterns of both humans and animals. Marey, the French physiologist, used a photographic "gun" in 1885 to record displacements in human gait and chronophotographic equipment to get a stick diagram of a runner. About the same time Muybridge in the United States sequentially triggered 24 cameras to record the patterns of a running man. Progress has been rapid during this century, and we now can record and analyze everything from the gait of a child with cerebral palsy to the performance of an elite athlete.

The term used for these descriptions of human movement is kinematics. Kinematics are not concerned with the forces, either internal or external, that cause the movement, but rather with the details of the movement itself. A complete and accurate quantitative description of the simplest movement requires a huge volume of data and a large number of calculations, resulting in an enormous number of graphical plots. For example, to describe the movement of the lower limb in the sagittal plane during one stride can require up to 50 variables. This includes linear and angular displacements, velocities, and accelerations. It should be understood that any given analysis may use only a small fraction of the available kinematic variables. An assessment of a running broad jump, for example, may require only the velocity, angle, and height of the body's center of gravity at takeoff. On the other hand, the energy analysis of an amputee's gait may require almost all the kinematic variables that are available.

2.1 KINEMATIC CONVENTIONS

In order to keep track of all the kinematic variables it is important to establish a convention system. In the anatomical literature a definite convention has been established, and we can completely describe a movement using terms such as proximal, flexion, and anterior. It should be noted that these terms are all *relative*, that is, they describe the position of one limb relative to another. They do not give us any idea as to where we are in space. Thus if we wish to analyze movement relative to the ground or the direction of gravity we must establish an absolute spatial reference system. Such conventions are mandatory when imaging devices are used to record the movement. However, when instruments are attached to the body the data becomes relative, and we lose information about gravity and the direction of movement.

2.1.1   Absolute Spatial Reference System

Several spatial references systems have been proposed; however, the one that is utilized throughout the text is the one that has been accepted for use in human gait. The vertical direction is Y, the direction of progression (anterior-posterior) is X, and the sideways direction (medial-lateral) is Z. Figure 2.1 depicts this convention; the positive direction is as shown. Angles must also have a zero reference and a positive direction. Angles in the XY plane are measured from $0°$ in the X direction, with positive angles being counterclockwise. Similarly, in the YZ plane, angles start at $0°$ in the Y direction and increase positively counterclockwise. The convention for velocities and accelerations follows correctly if we maintain the spatial coordinate convention:

$\dot{x}$—velocity in the X direction, positive when X is increasing

$\dot{y}$—velocity in the Y direction, positive when Y is increasing

$\dot{z}$—velocity in the Z direction, positive when Z is increasing

$\ddot{x}$—acceleration in the X direction, positive when $\dot{x}$ is increasing

$\ddot{y}$—acceleration in the Y direction, positive when $\dot{y}$ is increasing

$\ddot{z}$—acceleration in the Z direction, positive when $\dot{z}$ is increasing

The same applies for angular velocities and angular accelerations. A counterclockwise angular increase is a positive angular velocity, ω. When ω is increasing we calculate a positive angular acceleration, α.

2.1.2   Total Description of a Body Segment in Space

The complete kinematics of any body segment requires 15 data variables, all of which may be changing with time:

(i) position (x,y,z) of center of mass of segment

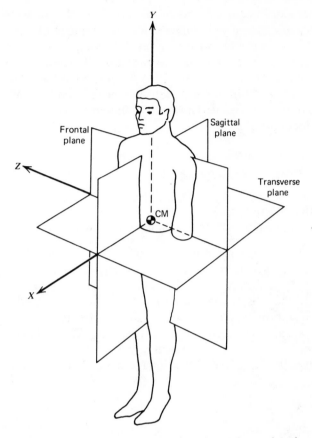

Figure 2.1   Spatial coordinate system used for all data and analyses.

(ii) linear velocity $(\dot{x}, \dot{y}, \dot{z})$ of center of mass of segment

(iii) linear acceleration $(\ddot{x}, \ddot{y}, \ddot{z})$ of center of mass of segment

(iv) angle of segment in two planes, θxy, θyz

(v) angular velocity of segment in two planes, ωxy, ωyz

(vi) angular acceleration of segment in two planes, αxy, αyz

Note that the third angle data is redundant; any segment's direction can be completely described in two planes. For a complete description of the total body (feet + legs + thighs + trunk + head + upper arms + forearms and hands = 12 segments) movement in three-dimensional space requires $15 \times 12 = 180$ data variables! It is no small wonder that we have yet to describe, let alone analyze, some of the more complex movements. Certain simplifications can certainly reduce the number of variables to a manageable number. In symmetrical level walking, for example, we can assume sagittal plane movement and can normally ignore the arm movement. The head, arms, and trunk (H.A.T.) are often considered to be a single segment, and with the symmetry we need to collect data from one lower limb only. The data variables in this case (four segments, one plane) can be reduced to 36.

2.2 DIRECT MEASUREMENT TECHNIQUES

2.2.1 Goniometers
(Representative paper—Finley and Karpovich, 1964)

A goniometer is a special name given to the electrical potentiometer that can be attached to measure a joint angle. As such, one arm of the goniometer is attached to one limb segment, the other to the adjacent limb segment and the axis of goniometer is aligned to the joint axis. In Figure 2.2 we see a fitting of a goniometer to a knee joint along with the equivalent electrical circuit. A constant voltage, E, is applied across the outside terminals, and the wiper arm moves to pick off a fraction of the total voltage. The fraction of the voltage depends on the joint angle, θ.

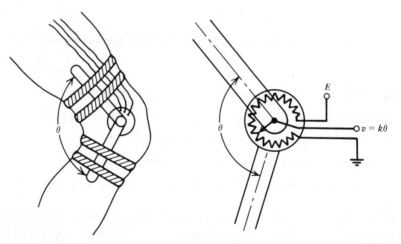

Figure 2.2   Mechanical and electrical arrangement of a goniometer located at the knee joint. Voltage output is proportional to the joint angle.

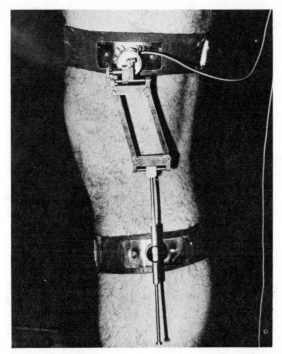

Figure 2.3 An electrogoniometer designed to accommodate changes in center of rotation of the knee joint, shown here fitted on a patient (By kind permission of Chedoke-McMaster Medical Center, Hamilton, Canada).

Thus the voltage on the wiper arm, $v = kE\theta = k_1\theta$ volts. Note that a voltage proportional to θ requires a potentiometer whose resistance varies linearly with θ. A goniometer designed for clinical studies is shown fitted on a patient in Figure 2.3.

Advantages

1. A goniometer is inexpensive.

2. Output signal is available immediately for recording or conversion into a computer.

Disadvantages

1. Relative angular data are given, not absolute angles, thus severely limiting diagnostic value.

2. It may require an excessive length of time to fit and align.

3. If a large number are fitted, movement can be encumbered by the straps and cables.

4. More complex goniometers are required for joints which do not move as hinge joints.

2.2.2 Accelerometers
(Representative paper—Morris, 1973)

As indicated by its name, an accelerometer is a device that measures acceleration. Most accelerometers are nothing more than force transducers designed to measure the reaction forces associated with a given acceleration. If the acceleration of a limb segment is a and the mass inside the accelerometer is m, then the force exerted by the mass is $F = ma$. This force is measured by a force transducer, usually a strain gauge or piezoresistive type. The mass is accelerated against a force transducer which produces a signal voltage, V, which is proportional to the force, and since m is known and constant, then V is also proportional to the acceleration. The acceleration can be toward or away from the face of the transducer; this is indicated by a reversal in sign of the signal. In most movements there is no guarantee that the acceleration vector will act at right angles to the face of the force transducer. The more likely situation is depicted in Figure 2.4, with the acceleration vector having a component normal to the transducer and another component tangent to the transducer face. Thus the accelerometer measures the a_n component. Nothing is known about a_t or a unless a triaxial accelerometer is used. Such a three-dimensional transducer is nothing more than three individual accelerometers mounted at right angles to each other, each one then reacting to the orthogonal component acting along its axis. Even with a triaxial accelerometer mounted on a limb there can be problems because of limb rotation, as indicated in Figure 2.5. In both cases the leg is accelerating in the same *absolute* direction, as indicated by vector a. The measured acceleration component, a_n, is quite different in each case. Thus the accelerometer is limited to those movements whose direction in space does not change drastically or to special contrived movements, such as horizontal flexion of the forearm about a fixed elbow joint.

A typical electrical circuit of a piezoresistive accelerometer is shown in Figure 2.6. It comprises a half bridge consisting of two equal resistors, R_1. Within the transducer resistors Ra and Rb change their resistances proportional to the acceleration acting against them. With no acceleration $Ra = Rb = R_1$ and with the balance potentiometer properly adjusted the voltage at terminal 1 is the same as at terminal 2; thus the output voltage is $V = 0$. With the acceleration in the direction shown, Rb increases and Ra

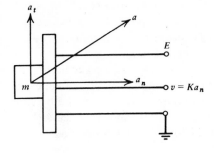

Figure 2.4 Schematic of an accelerometer, showing acceleration with normal and tangential components. Voltage output is proportional to the normal component of acceleration, a_n.

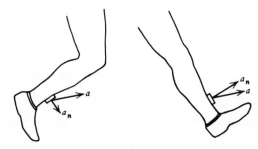

Figure 2.5 Two movement situations where acceleration in space is identical yet the normal component is quite different.

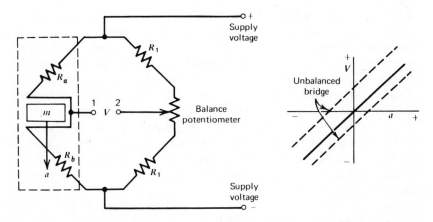

Figure 2.6 Electrical bridge circuit used in most force transducers and accelerometers. See text for detailed operation.

decreases; thus the voltage at terminal 1 increases. The resultant unbalance in the bridge circuit results in voltage, V, proportional to the acceleration. Conversely, if the acceleration is upward, Rb decreases and Ra increases; the bridge unbalances in the reverse direction, giving a signal of the opposite polarity. Thus, over the dynamic range of the accelerometer the signal is proportional to both the magnitude and direction of acceleration acting along the axis of the accelerometer. However, if the balance potentiometer is not properly set we have an unbalanced bridge and we could get a voltage/acceleration relationship like that indicated by the dotted lines.

Advantages

1. Output signal is available immediately for recording or conversion into a computer.

Disadvantages

1. Acceleration signal is *relative* to its position on the limb segment.

2. Cost can be excessive if a large number are used.

3. If a large number are used they can encumber movement.

4. Many types are quite sensitive to shock and are easily broken.

2.3 IMAGING MEASUREMENT TECHNIQUES

The Chinese proverb "a picture is worth more than ten thousand words" holds an important message for any human observer, including the biomechanics researcher interested in human movement. Because of the complexity of most movements the only system that can possibly capture all the data is an imaging system. Given the additional task of describing a dynamic activity we are further challenged in having to capture data over an extended period of time. This necessitates taking many images at regular intervals during the event.

There are many types of imaging systems that could be used; however, the discussion will be limited to four different types—movie camera, television, multiple exposure, and optoelectric types. Whichever system is chosen a lens is involved; therefore a short review of basic optics is given here.

2.3.1 Review of Basic Lens Optics

A simple converging lens is one which creates an inverted image in focus at a distance v from the lens. As seen in Figure 2.7, if the lens-object

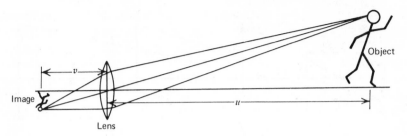

Figure 2.7   Simple focusing lens system showing relationship between object and image.

distance is u then the focal length of the lens, f, is as follows:

$$\frac{1}{f} = \frac{1}{v} + \frac{1}{u}. \qquad (2.1)$$

The imaging systems used for movement studies are such that the object-lens distance is quite large compared with the lens-image distance. Therefore

$$\frac{1}{u} \approx 0, \frac{1}{f} = \frac{1}{v} \text{ or } f = v.$$

Thus if we know the focal length of the lens system we can see that the image size is related to the object size by a simple triangulation. A typical focal length is 25 mm, a wide angle lens is 13 mm, and a telephoto lens is 150 mm. A zoom lens is just one in which the focal length is infinitely variable over a given range. Thus as L increases the focal length must increase proportionately to produce the same image size. Figure 2.8 illustrates this principle. For maximum accuracy it is highly desirable that the image be as large as possible. Thus it is advantageous to have a zoom lens rather than a series of fixed lenses; individual adjustments can be readily made for each movement to be studied, or even during the course of the event.

2.3.2 *f*-Stop Setting and Field of Focus

The amount of light entering the lens is controlled by the lens opening, which is measured by its *f*-stop setting (*f* means fraction of lens aperture opening). The larger the opening the lower the *f*-stop setting. Each *f*-stop setting corresponds to a proportional change in the amount of light allowed in. A lens may have the following settings: 22, 16, 11, 8, 5.6, 4, 2.8, and 2, *f*/22 is 1/22 of the lens diameter and *f*/11 is 1/11 of the lens diameter. Thus *f*/11 lets in four times the light that *f*/22 does. The fractions are

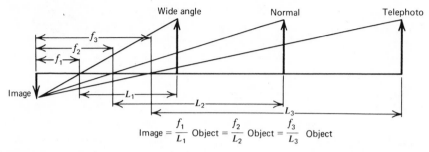

Figure 2.8   Differences in focal length of wide angle, normal, and telephoto lenses result in same image size.

arranged so that each one lets in twice the light of the adjacent higher setting (e.g., $f/2.8$ is twice the light of $f/4$).

To keep the lighting requirements to a minimum it is obvious that the lens should be opened as wide as possible to a low f setting. However, problems occur with the field of focus. This is defined as the range of the object that will produce a focused image. The lower the f setting, the narrower the range over which an object will be in focus. For example, if we wish to photograph a movement which was to move over a range from 10 to 30 ft we could not reduce the f-stop below 5.6. The range set on the lens would be about 15 feet, and everything between 10 and 30 ft would remain in focus. The final decision regarding f-stop depends on the movie camera shutter speed and film speed.

2.3.3   Cinematography
(Representative paper—Eberhart and Inman, 1951)

Many different sizes of movie cameras are available. 8mm cameras are the smallest (they actually use 16mm film which is run through the camera twice, then split into two 8-mm strips after it is developed); then come 16mm, 35mm, and 70mm. The image size of 8mm is somewhat small for accurate measurements, while 35mm and 70mm movie cameras are too expensive to buy and to operate. Thus 16mm cameras have evolved as a reasonable compromise, and most high speed movie cameras are 16mm.

There are several types of 16mm cameras available. Some are spring driven; others are motor driven by either batteries or power supplies from alternating current sources. Battery-driven types have the advantage of being portable to sites where power is not available.

The type of film required depends on the lighting available. The ASA rating is a measure of the "speed" of the film; the higher the rating, the less light is required to get the same exposure. 4-X Reversal film with an ASA rating of 400 is a common type. Higher ASA ratings are available and are good for a qualitative assessment of movement, especially fast-moving sporting events. However, the coarse grain of these higher ASA films introduces inaccuracies in quantitative analyses.

The final factor that influences the lighting required is the shutter speed of the camera. The higher the frame rate, the less time is available to expose the film. Most high-speed cameras have rotating shutters that open once per revolution for a period of time to expose a new frame of unexposed film. The arc of the opening, as depicted in Figure 2.9, and the speed of rotation of the shutter decide the exposure time. For example, at 60 frames/sec using a 3 factor shutter the exposure time is 1/180 sec. The amount of light entering will be the same as a normal (still) camera set to a speed of 1/180 sec.

To make the final settings we use an exposure meter to measure the light intensity on the human subject. For a given filming the variables that are preset are film ASA, shutter factor, and frame rate. The frame rate is set low enough to capture the desired event, but not too high to require extra lighting or result in film wastage. To understand the problem associated with the selection of an optimal rate, the student is referred to Section 2.5.3 on the Sampling Theorem. The final variable to decide is the f-stop. The light meter gives an electrical meter reading proportional to

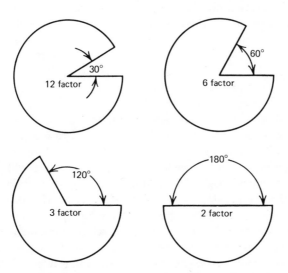

Figure 2.9 Various factor shutters used in movie cameras. Film is exposed during the opening arc and is advanced while the shutter is closed.

the light intensity, such that when the film ASA and exposure time are set the correct f-stop can be determined. Thus with the movie camera set at the right frame rate, f-stop, and range, the filming is ready to commence.

2.3.4 Television
(Representative paper—Winter et al., 1972)

The major difference between television and cinematography is the fact that television has a fixed frame rate. The name given to each television image is a field. In North America there are 60 fields/sec, in Europe the standard is 50 fields/sec. Thus television has a high enough field rate for most movements, but probably too low for a quantitative analysis of rapid athletic events. The f-stop, focus, and lighting for television can be adjusted by watching the television monitor as the controls are varied. Many television cameras have electronic as well as optical controls which influence brightness and contrast. Also, focus can be adjusted electronically as well as optically. The major advantage of television is the capability for instant replay, which serves both as a quality control check and as an initial qualitative assessment.

2.3.5 Multiple Exposure
(Representative paper—Murray et al, 1964)

One of the most economical and simplest imaging systems is the multiple exposure technique. It uses a still camera in a darkened room with its shutter open for the entire time of the event. A bright flashing light, called a strobe light, illuminates the subject for a very short time (a few milliseconds); this is repeated at regular intervals (say, 20 times/sec). Reflective strips or markers are placed on the subject, and once every flash a new exposure is made on the film. Over a period of a few seconds the film is re-exposed many times, but each time with the body in a new position. Thus the final exposure resembles a "stick" diagram of the subject with the position of the limbs shown at equal intervals of time. A flashing rate of 20/sec is equivalent to a movie camera rate of 20 frames/sec. In Figure 2.10 we see a typical image of a walking subject. Note that the background must be darkened and that there is considerable overlap of images, especially when the movement is slow. Such a technique is not only inexpensive but also has the advantage of giving rapid assessment if a Polaroid type camera is used. However, because of the overlaying of the images on top of each other, the assessments are limited to crude measures such as range of movement. Because of the need for a darkened room and the

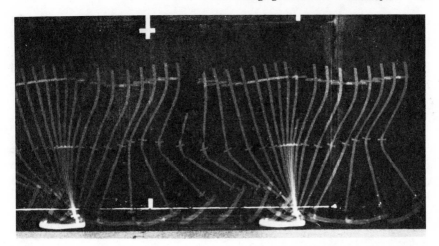

Figure 2.10 Multiple exposure image of one and one-half strides during walking. Subject wears darkened clothes with reflective tape on each limb. Strobe light was flashed at 20 Hz in a darkened room.

limited field of view of a fixed camera, multiple techniques have restricted applications. Also, the flashing strobe light can be very distracting to the subject.

2.3.6 Optoelectric Techniques
(Representative paper—Öberg, 1974)

In the past few years there have been several new developments in optoelectric imaging systems that have certain advantages over cinematography and television. Two types of systems have evolved. The first is a commercial development called Selspot, which requires the subject to wear special lights on each desired anatomical landmark. The lights are flashed sequentially and the location (x,y) of the light flash is picked up on a special camera. The camera consists of a standard lens system focusing the light flash on a special semi-conductor diode surface. The location of the image of the light flash gives two signals, one indicating the x coordinate of the image, the other indicating the y coordinate. As each light flashes in sequence a series of x- and y-coordinate signals are fed to a tape recorder or high speed computer.

The second type is in effect a coarse television system made up of a matrix of light-sensitive diodes. Special electronic circuits scan the x and y connections to each diode. If a diode is illuminated its x and y

coordinates are fed to a high speed computer. By sequentially scanning all the coordinates of the matrix at regular intervals a temporal history of x and y coordinates is obtained for each marker. The identification and labeling of each marker must be left to the software programs developed for the computer.

2.3.7   Summary of Various Kinematic Systems

It is useful to tabulate a summary of pros and cons of each of the measurement techniques. This is presented in Table 2.1; it should be cautioned that the importance of each advantage or disadvantage must be tempered by the needs and constraints of each particular application.

2.4   DATA CONVERSION TECHNIQUES

2.4.1   Analog-to-Digital Convertors

To students not familiar with electronics the process that takes place during conversion of a physiological signal into a digital computer can be somewhat mystifying. A short schematic description of that process will now be given. An electrical signal representing a force, an acceleration, an EMG potential, or the like, is fed into the input terminals of the analog-to-digital convertor. The computer controls the rate at which the signal is sampled; the optimal rate is governed by the sampling theorem (see Section 2.5.3).

Figure 2.11 depicts the various stages in the conversion process. The first is a sample/hold circuit in which the analog input signal is changed into a series of short duration pulses, each one equal in amplitude to the original analog signal at the time of sampling (these times are specified by the computer operator). The final stage of conversion is to translate the amplitude and polarity of the sampled pulse into digital format. This is usually a binary code in which the signal is represented by a number of bits. For example, a twelve bit code represents $2^{12} = 4096$ levels. This means that the original sampled analog signal can be broken into 4096 discrete amplitude levels, with a unique code representing each of these levels. Each coded sample (consisting of 0's and 1's) forms a 12 bit "word," which is rapidly stored in computer memory for recall at a later time. If a 5-sec signal were converted at a sampling rate of 100 times per second there would then be 500 data words stored in memory to represent the original 5-sec signal.

Table 2.1 Summary of Advantages and Disadvantages of Common Kinematic Systems

Considerations	Goniometers	Accelerometers	Cinematography	Television	Multiple Exposure
Cost Capital + Running	Low, except cost of pen or tape recorders	Expensive, including cost of electronics and recorders	Moderate, except cost of conversion equipment	Moderate, cost of conversion is high	Low
Encumberment to gait	Can encumber if many are used	Can restrict movement if many are used	Minimal	Minimal	Minimal
Time to attach and calibrate	Can be excessive to attach and calibrate	Can be excessive to attach	Minimal to attach markers	Minimal to attach markers	Minimal to attach markers
Availability of Data for Analysis	Instantly available	Instantly available	Development of film and conversion of data can be high	Instant replay, has capability for instant conversion	Instantly available
Data Format	Assumes hinge joint relative angles only	Absolute direction of acceleration not known	Absolute coordinates	Absolute coordinates, maximum field rate of 60 Hz	Overlapping of exposures limits accuracy and frame rate
Other Considerations	Unless telemetry is used, range and speed of activity are restricted	Unless telemetry is used, range and speed of activity are restricted	Extra lighting required indoors; permanent visual record available	Extra lighting required	Low flashing rates are distracting, darkened room required

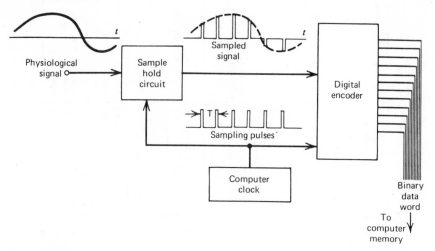

Figure 2.11   Schematic diagram showing the steps involved in an analog-to-digital conversion of a physiological signal.

2.4.2   Movie Conversion Techniques

As 16mm movie cameras are the most common form of data collection, it is important to be aware of various coordinate extraction techniques. Each system that has evolved requires the projection of each movie frame on some form of screen. The most common type requires the operator to move a mechanical x-y coordinate system until a pointer, light, or crosshair lies over the desired anatomical landmark. Then the x and y coordinates can be read off or transferred to a computer at the push of a button. Figure 2.12 shows the component parts of such a conversion system.

A second type of system involves the projection of the film image onto a special grid system. When the operator touches the grid with a special pen the coordinates are automatically transferred into a computer. Both systems are limited to the speed and accuracy of the human operator. Our experience indicates that an experienced operator can convert an average of 15 coordinate pairs per minute. Thus a 3-sec film record filmed at 50 frames/sec could have five markers converted in 50 min.

2.4.3   Television Conversion

Television appears on the surface to be the answer to the problem of automation because the image is already an electronic signal. The prob-

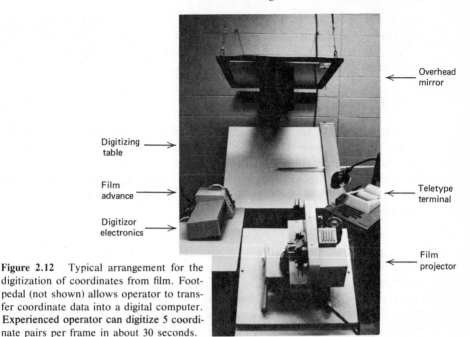

Digitizing table →

Film advance →

Digitizor electronics →

← Overhead mirror

← Teletype terminal

← Film projector

Figure 2.12   Typical arrangement for the digitization of coordinates from film. Foot-pedal (not shown) allows operator to transfer coordinate data into a digital computer. Experienced operator can digitize 5 coordinate pairs per frame in about 30 seconds.

lem arises in the conversion of such a high-frequency signal. One line across the television screen is scanned in about 60 μsec, and in this short time the full width of the image is scanned. If the image width represents 120 cm across the subject then each μsec of sweep time represents 2 cm. Therefore, in order to sample every .5 cm the sweep signal must be sampled every $\frac{1}{4}$ μsec. This requires an analog-to-digital convertor capable of sampling at a 4-MHz rate. A flexible television conversion interface has been developed (Dinn et al, 1970) and has been extensively used for locomotion studies (Winter and colleagues, 1972). This interface requires a reasonably high-speed digital computer. However, the advantage is that coordinate data are converted in real time, without any human error.

2.5   PROCESSING OF RAW KINEMATIC DATA

2.5.1   Nature of Unprocessed Data

The converted coordinate data from film or television is called "raw" data. This means that it has additive noise from many sources—vibration in the movie camera and projector or random errors due to the conversion process (human errors in determining the center of a body marker and precision of the conversion device itself). In order to understand the techniques used to smooth the data an appreciation of harmonic (or frequency) analysis is necessary.

2.5.2   Harmonic (Fourier) Analysis

(a) *Alternating Signals.*   An alternating signal (often called ac, for alternating current) is one that is continuously changing with time. It may be periodic or completely random or a combination of both. Also, any signal may have a dc (direct current) component, which may be defined as the bias value about which the ac component fluctuates. Figure 2.13 shows example signals.

(a) *Frequency Content.*   Any of these signals can also be discussed in terms of its frequency content. A sine (or cosine) waveform is a single frequency; any other waveform can be the sum of a number of sine waves. Several names are given to the graph showing these frequency components—spectral plots, harmonic plots, or spectral density functions. Each shows the amplitude or power of each frequency component plotted against frequency; the mathematical process to accomplish this is called a Fourier transformation or Harmonic analysis. In Figure 2.14 are plotted time domain signals and their equivalent in the frequency domain.

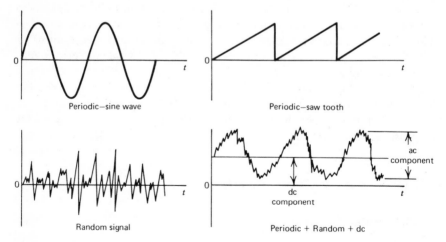

Figure 2.13   Time-related waveforms to demonstrate characteristics of biomechanical signals that may require processing.

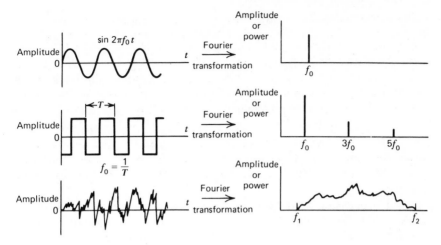

Figure 2.14 Relationship between a signal as seen in the time domain and its equivalent in the frequency domain.

Note that the Fourier transformation of periodic signals has discrete frequencies, while nonperiodic signals have a continuous spectrum, defined by its lowest frequency f_1, and its highest frequency, f_2. To analyze a periodic signal we must express the frequency content in multiples of the fundamental, f_0. These higher discrete frequencies are called harmonics. The third harmonic is $3f_0$, and tenth harmonic is $10f_0$. Any periodic signal can be broken down into its harmonic components. The sum of the proper amplitude of these harmonics is called a Fourier series.

Thus a given signal, $V(t)$, can be expressed as:

$$V(t) = Vdc + V_1 \sin (\omega t + \theta_1) \qquad (2.2)$$
$$+ V_2 \sin (2\omega t + \theta_2) + \cdots Vn \sin (n\omega t + \theta n)$$

where: $\omega = 2\pi f_0$.

For example, a square wave of amplitude V can be described by the Fourier series:

$$V(t) = \frac{4V}{\pi} \left(\sin \omega t + \frac{1}{3} \sin 3\omega t + \frac{1}{5} \sin 5\omega t + \cdots\right) \qquad (2.3)$$

A triangular wave of duration $2t$ and repeating itself every T seconds is:

$$V(t) = \frac{2Vt}{T} \left(\tfrac{1}{2} + \left(\frac{2}{\pi}\right)^2 \cos \omega t + \left(\frac{2}{3\pi}\right)^2 \cos 3\omega t + \cdots\right) \qquad (2.4)$$

2.5.3 Sampling Theorem

Film and television are sampling processes. They capture the move-ment event for a short period of time, after which no further changes are recorded until the next field or frame. Playing a movie film back slowly demonstrates this phenomenon: the image jumps from one position to the next in a distinct step rather than a continuous process. The only reason film or television does not appear to jump at normal projection speeds (24/sec for film, 60/sec for television) is because the eye can retain an image for a period of about 1/15 sec. The eye's short-term "memory" enables the human observer to average or smooth out the jumping move-ment.

In the processing of any time-varying data, no matter what its source, the sampling theorem must not be violated. Without going into the math-ematics of sampling process the theorem states that "the process signal must be sampled at a frequency at least twice as high as the highest frequency present in the signal itself." If we sample a signal at too low a frequency we can get aliasing errors. This results in false frequencies, frequencies that were not present in the original signal, being generated in the sample data. Refer to Figure 2.15 to see this effect. Both signals are being sampled at the same interval, T. Signal 1 is being sampled about ten times per cycle, while signal 2 is being sampled less than two times per cycle. Note that the amplitude of the samples taken from signal 2 are

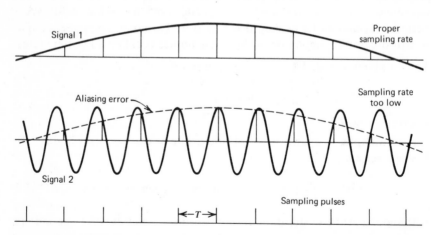

Figure 2.15 Sampling of two signals, one at a proper rate, the other at too low a rate. Signal 2 is sampled at a rate less than twice its frequency, such that its sampled amplitudes are the same as for Signal 1. This represents a violation of the sampling theorem and results in an error called aliasing.

identical to those sampled from signal 1. A false set of sampled data has been generated from signal 2 because the sample rate is too low—the sampling theorem has been violated.

Possibly the major fault in filming is not too low a rate. The tendency is to play it safe and film at too high a rate. Usually there is a cost associated with such a decision. The initial cost is probably in the equipment required. A high-speed movie camera can cost four or five times as much as a standard (24 frames/sec) model. Or a special optoelectric system complete with the necessary computer can be a $50,000 decision. In addition to these capital costs there are the higher operational costs of converting the data and running the necessary kinematic and kinetic computer programs. Except for higher-speed running and athletic movements it is quite adequate to use a standard movie or television camera. For normal and pathological gait studies it has been shown that kinetic and energy analyses can be done with negligible error using a standard 24 frame/sec movie camera (Winter and Wells, 1978).

2.5.4   Signal versus Noise

In the study of movement the signal may be an anatomical coordinate that changes with time. For example, in walking, the Y (vertical) coordinate of the heel will have certain frequencies which will be higher than those associated with vertical coordinate of the knee or trunk. Similarly, the frequency content of all trajectories will increase in running compared with walking. In a repetitive movement such as walking the frequencies present will be multiples (harmonics) of the fundamental frequency (stride frequency). When walking at 120 steps/min (2 Hz) the stride frequency is 1 Hz. Therefore we can expect to find harmonics at 2 Hz, 3 Hz, 4 Hz, and so on. Normal walking has been analyzed by digital computer and the harmonic content of the trajectories of seven leg and foot markers determined (Winter et al., 1974). The highest harmonics were found to be in the toe and heel trajectories, and it was determined that 99.7% of the signal power was contained in the lower seven harmonics (below 6 Hz). The harmonic analysis for the heel marker for 20 subjects is shown in Figure 2.16.

Above the seventh harmonic there was still some signal power, but it had the characteristics of "noise." Noise is the term used to describe components of the final signal that is not due to the process itself (in this case, walking). Noise can be introduced by the measurement process: vibrations in the movie camera or imperfect alignment of the film in the sprockets; or by the conversion process: human and machine errors in the extraction of coordinates from the film. If the total effect of all these

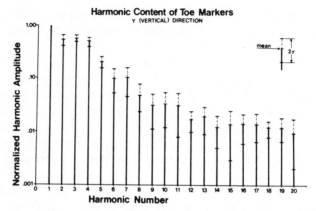

Figure 2.16   Harmonic content of the vertical displacement of a toe marker from 12 subjects during normal walking. Fundamental frequency (harmonic number = 1) is normalized at 1.00. Over 99% of power is contained below the 7th harmonic. (By permission of the *Journal of Biomechanics.*)

errors is random, then the true signal will have an added random component. Usually the random component is high frequency, as is borne out in Figure 2.16. Here we see evidence of higher-frequency components extending up to the twentieth harmonic, which was the highest frequency analyzed. The presence of the higher-frequency noise is of considerable importance when we consider the problem of trying to calculate velocities and accelerations. Consider the process of time differentiation of a signal containing additive higher-frequency noise. Suppose the signal can be represented by a summation of harmonics:

$$x = \sum_{n=1}^{\Lambda} X_n \sin (n\omega_0 t + \theta_n) \qquad (2.5)$$

where: ω_0 is the fundamental frequency.
n is the harmonic number.
X_n is the amplitude of the nth harmonic.
θ_n is the phase of the nth harmonic.

To get the velocity in the x direction, Vx, we differentiate with respect to time:

$$Vx = \frac{dx}{dt} = \sum_{n=1}^{\Lambda} n\omega_0 X_n \cos (n\omega_0 t + \theta_n) \qquad (2.6)$$

Similarly, the acceleration, Ax, is:

$$Ax = \frac{dVx}{dt} = -\sum_{n=1}^{\Lambda} (n\omega_0)^2 X_n \sin (n\omega_0 t + \theta_n) \qquad (2.7)$$

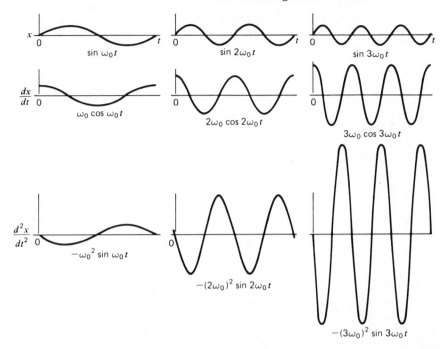

Figure 2.17   Relative amplitude changes as a result of time differentiation of signals of increasing frequency. First derivative increases amplitude proportional to frequency; second derivative increases amplitude proportional to (frequency)^2.

Thus the amplitude of each of the harmonics increases with its harmonic number; for velocities they increase linearly, and for accelerations the increase is proportional to the square of the harmonic number. This phenomenon is demonstrated in Figure 2.17 where the fundamental, second, and third harmonics are shown along with their first and second time derivatives. Assuming that the amplitude, x, of all three components is the same, we can see that the first derivative (velocity) of harmonics increases linearly with increasing frequency. The first derivative of the third harmonic is now three times that of the fundamental. For the second time derivative the increase repeats itself and the third harmonic acceleration is now nine times that of the fundamental.

In the trajectory data for gait, x_1 might be 5 cm and $x_{20} = .5$ mm. The twentieth harmonic noise is hardly perceptible in the displacement plot. In the velocity calculation the twentieth harmonic increases 20-fold so that it is now one-fifth that of the fundamental. In the acceleration calculation the twentieth harmonic increases another factor of 20 and now is four times the magnitude of the fundamental. This effect is shown in Figure

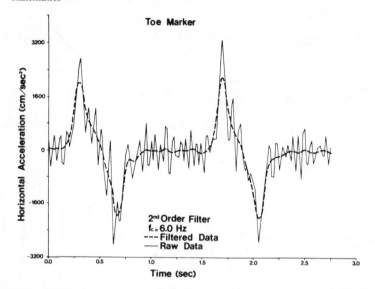

Figure 2.18   Horizontal acceleration of the toe marker during normal walking as calculated from displacement data from television. Solid line is acceleration based on unprocessed "raw" data; dotted line is that calculated after data are filtered twice with a second order, low-pass digital filter (By permission of the *Journal of Biomechanics*).

2.18, which plots the acceleration of the toe during walking. The random-looking signal is the raw data differentiated twice. The smooth signal is the acceleration calculated after most of the higher-frequency noise has been removed. Techniques to accomplish this now will be discussed.

2.5.5   Smoothing and Fitting of Data

The removal of noise can be accomplished several ways. The aims of each technique are basically the same. However, the results do differ somewhat.

(a) *Curve Fitting Techniques.*   The basic assumption here is that the trajectory signal has a predetermined shape and that by fitting the assumed shape to a "best fit" with the raw data a smooth signal will result. For example, it may be assumed that the data is a certain order polynomial:

$$x(t) = a_0 + a_1 t + a_2 t^2 + a_3 t^3 + \cdots a_n t^n \qquad (2.8)$$

By computer techniques, the coefficients $a_0 \cdots a_n$ can be selected to give a best fit, using such criteria as minimum mean square error. The complexity of the curve fitting can be quite restrictive in terms of computer time. A similar fit can be made assuming a certain number of harmonics:

$$x(t) = a_0 + \sum_{n=1}^{N} a_n \sin (n\omega_0 t + \theta_n) \tag{2.9}$$

This model has a better basis especially in repetitive movement, while the polynomial may be better in certain nonrepetitive movement, such as broad jumping. A third technique, spline curve fitting, is a modification of the polynomial technique. The curve to be fitted is broken into sections, each section starting and ending with an inflection point, with special fitting being done between adjacent sections. The major problem with this technique, other than computer time, is the error introduced by improper selection of the inflection points. These inflection points must be determined from the noisy data, and as such are strongly influenced by the very noise that we are trying to eliminate.

(b) *Digital Filtering.*   Recent technological advances in digital filtering have opened up a much more promising and less restrictive solution to the noise reduction. The basic approach can be described by analyzing the frequency spectrum of both signal and noise. Figure 2.19a shows a schematic plot of a signal and noise spectrum. As can be seen, the signal is assumed to occupy the lower end of the frequency spectrum and overlaps with the noise, which is usually higher frequency. Filtering of any signal is aimed at the selective rejection, or attenuation, of certain frequencies. In the above case the obvious filter is one which passes, unattenuated, the lower frequency signals, while at the same time attenuating the higher-frequency noise. Such a filter, called a low-pass filter, has a frequency response as shown in Figure 2.19b. The frequency response of the filter is the ratio of the output of the filter, $X_0(f)$, to the input of the filter, $X_i(f)$, at each frequency present. As can be seen the response at lower frequencies is 1.0. This means that the input signal passes through the filter unattenuated. However, there is a sharp transition at the cutoff frequency f_c, so that the signals above f_c are severely attenuated. The net result of the filtering process can be seen by plotting the spectrum of the output signal, $X_0(f)$, as seen in Figure 2.19c. Two things should be noted, First, the higher-frequency noise has been severely reduced but not completely rejected. Second, the signal, especially in the region where the signal and noise overlap (usually around f_c), is also slightly attenuated. This results in a slight distortion of the signal. Thus a compromise has to be made in the selection of the cutoff frequency. If f_c is set too high, less signal distortion occurs but too much noise is allowed to pass. Conversely, if f_c is too low, the noise is drastically reduced, but at the expense of increased

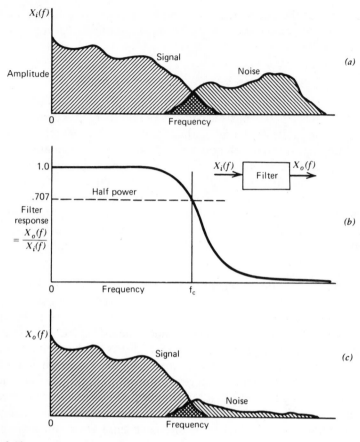

Figure 2.19
a Hypothetical frequency spectrum of a waveform, consisting of the desired frequency signal and the unwanted higher frequency noise.
b Response of low-pass filter, $X_o(f)/X_i(f)$, used to reduce the noise component.
c Spectrum of the output waveform, obtained by multiplying the amplitude of the input by the filter response at each frequency. Higher-frequency noise is severely attenuated, while the signal is passed with only minor distortion (in the transition region around cut-off frequency).

signal distortion. A sharper cutoff filter will improve matters, but at an additional expense. In digital filtering this means a more complex digital filter, and thus more computer time.

The theory behind digital filtering (Radar and Gold, 1967) will not be covered, but the application of low-pass digital filtering to kinematic data processing will be described in detail. First, it must be assessed what the signal spectrum is as opposed to the noise spectrum. This can readily be done, as is seen in the harmonic analysis presented previously in Figure 2.16. As a result of the previous discussion the cutoff frequency of a digital filter should be set at about 6 Hz. The format of a digital filter, which processes the film data in the time domain, is as follows:

$$X^1(nT) = a_0 X(nT) + a_1 X(nT-T) + a_2 X(nT-2T) + b_1 X^1(nT-T)$$
$$+ b_2 X^1(nT-2T) \qquad (2.10)$$

where: X^1 refers to filtered output coordinates.
X refers to unfiltered coordinate data.
nT is the nth sample frame.
$(nT-T)$ is the $(n-1)$th sample frame.
$(nT-2T)$ is the $(n-2)$th sample frame.
$a_0, \cdots, b_2$, etc. are the filter coefficients.

These filter coefficients are constants that depend on the type and order of the filter, the sampling frequency, and the cutoff frequency. As can be seen, the filtered output, $X^1(nT)$, is a weighted version of the immediate and past raw data plus a weighted contribution of past filtered output.

The order of the filter decides the sharpness of the cutoff. The higher the order the sharper the cutoff, but the larger the number of coefficients. For example, a Butterworth-type low-pass filter, of second order, is to be designed to cut off at 6 Hz, using film data taken at 60 Hz (60 frames/sec). All that is required to determine these coefficients is the ratio of the sampling frequency to cutoff frequency. In this case it is 10. The design of such a filter would yield the following coefficients: $a_0 = .06745$, $a_1 = .1349$, $a_2 = .06745$, $b_1 = 1.1430$, $b_2 = -.4128$.

Note that the algebraic sum of all the coefficients = 1.0000; this gives a response of unity over the pass band. In Table 2.2 are listed the coefficients for use in a second-order Butterworth filter for various values of fs/fc. Note that the same filter coefficients could be used in many different applications, as long as the ratio fs/fc is the same. For example, an EMG signal sampled at 2000 Hz with cutoff desired at 400 Hz would have the same coefficients as one employed for movie film coordinates where the film rate was 30 Hz and cutoff was 6 Hz.

As well as attenuating the signal there is a phase shift of the output signal relative to the input. For this second-order filter there is a 90° phase lag at the cutoff frequency. This will cause a second form of distortion, called phase distortion, to the higher harmonics within the band pass region. Even more phase distortion will occur to those harmonics above fc, but these components are mainly noise and are being severely attenuated. This phase distortion may be more serious than the amplitude distortion that occurs to the signal in the transition region. To cancel out

Table 2.2   Coefficients of Second-Order Low-Pass, Digital Filter

fs/fc	a_0	a_1	d_2	b_1	b_2
4.0	.2929	.5858	.2929	0	-.1716
5.0	.2066	.4132	.2066	.3695	-.1959
6.0	.15505	.3101	.15505	.6202	-.2404
7.0	.1212	.2424	.1212	.8030	-.2878
8.5	.0884	.1768	.0884	1.0011	-.3547
10.0	.06745	.1349	.06745	1.1430	-.4128
12.0	.0495	.0990	.0495	1.2796	-.4776
14.0	.0379	.0758	.0379	1.3789	-.5305
16.0	.02995	.0599	.02995	1.4542	-.5740
18.0	.0243	.0486	.0243	1.5134	-.6106
20.0	.0201	.0402	.0201	1.5610	-.6414

this phase lag the once-filtered data can be filtered again, but this time in the reverse direction of time. This introduces an equal and opposite phase lead so that the net phase shift is zero. Also, the cutoff of the filter will be twice as sharp as for single filtering. In effect, by this second filtering in the reverse direction, we have created a fourth-order, zero phase shift filter, which yields a filtered signal that is back in phase with the raw data but with most of the noise removed.

In Figure 2.20 we see the frequency response of a second-order Butter-

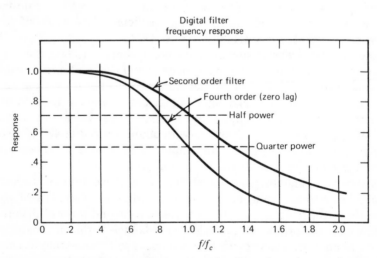

Figure 2.20   Response of a second-order, low-pass filter. Curve is normalized to 1.0 at cut-off frequency, F_c. Because of phase lag of filter, a second filtering is done in the reverse direction of time, which results in a fourth order, zero-lag, filter.

worth filter, normalized with respect to the cutoff frequency. Superimposed on this curve is the response of the fourth-order, zero phase shift filter. Thus the new cutoff frequency is .802 of the original cutoff frequency; this will have to be taken into account when specifying fc for the original filter. If the desired cutoff frequency is 6 Hz, the original second-order filter will have to have a cutoff of $6.0 \div .802 = 7.48$ Hz.

The application of one of these filters in smoothing raw coordinate data can now be seen by examining the data that yielded the harmonic plot of Figure 2.16. The horizontal acceleration of this toe marker, as calculated by finite differences from the filtered data, is plotted in Figure 2.18. Note how repetitive the filtered acceleration is and how it passes through the "middle" of the noisy curve, as calculated using the unfiltered data. Also, note that there is no phase lag in this filtered data because of the dual forward and reverse filtering processes.

2.5.6   Comparison of Some Smoothing Techniques

It is valuable to see the effect of several different curve-fitting techniques on the same set of noisy data. The following summary of a validation experiment which was conducted to compare (Pezzack et al., 1977) three commonly used techniques illustrates the wide differences in the calculated accelerations.

Data obtained from the horizontal movement of a lever arm about a vertical axis were recorded three different ways. A goniometer on the axis recorded angular position, an accelerometer mounted at the end of the arm gave tangential acceleration and thus angular acceleration, and cinefilm data gave image information which could be compared with the angular and acceleration records. The comparisons are given in Figures 2.21a, b, c, and d. Figure 2.21a compares the angular position of the lever arm as it was moved from rest through about 130° and back to the original position. The goniometer signal and the lever angle as analyzed from the film data are plotted, and compare very closely. The only difference is that the goniometer record is somewhat noisy compared with the film data.

Figure 2.21b compares the directly recorded angular acceleration, which can be calculated by dividing the tangential acceleration by the radius of the accelerometer from the center of the rotation, with the angular acceleration as calculated via the second derivative of the digitally filtered coordinate data (Winter et al., 1974). The two curves match extremely well, and the finite difference acceleration exhibits less noise than the directly recorded acceleration.

Figure 2.21c compares the directly recorded acceleration with the

38   Kinematics

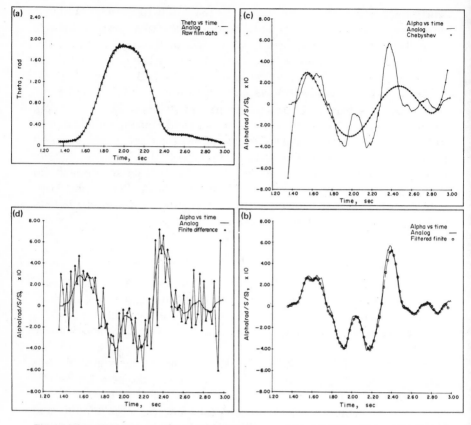

Figure 2.21   Comparison of several techniques used to determine the acceleration of movement based on film displacement data.

a Displacement angle of a simple extension/flexion, as plotted from film and goniometer data.

b Acceleration of movement in *a*, as measured by accelerometer and as calculated from film coordinates after digital filtering.

c Acceleration as determined from a 9th order polynomial fit of the displacement data compared with directly recorded acceleration.

d Acceleration as determined by finite difference techniques compared with measured curve. (By permission of *Journal of Biomechanics*.)

calculated angular acceleration using a polynomial fit on the raw angular data. A ninth order polynomial was fitted to the angular displacement curve to yield the following fit:

$$\theta(t) = .064 + 2.0t - 35t^2 + 210t^3 - 430t^4 + 400t^5$$
$$- 170t^6 + 25t^7 + 2.2t^8 - .41t^9 \text{ rad} \qquad (2.11)$$

Note that θ is in radians and t in seconds. To get the curve for angular acceleration all we need do is take the second time derivative to yield:

$$\alpha(t) = -70 + 1260t - 5160t^2 + 8000t^3 + 5100t^4$$
$$+ 1050t^5 + 123t^6 - 29.5t^7 \text{ rad/sec}^2 \qquad (2.12)$$

This acceleration curve compared with the accelerometer signal shows considerable discrepancy, enough to cast considerable doubt on the value of the polynomial fit technique. The polynomial is fitted to the displacement data in order to get an analytic curve which can be differentiated to yield another smooth curve. Unfortunately, it appears that a considerably higher-order polynomial would be required to achieve even a crude fit; however, the computer time might become too prohibitive.

Finally, in Figure 2.21d we see the accelerometer signal plotted against angular acceleration as calculated by second-order finite difference techniques. The plot speaks for itself—the accelerations are too noisy to mean anything.

2.6 CALCULATION OF ANGLES FROM SMOOTHED DATA

2.6.1 Limb Segment Angles

Given the coordinate data from anatomical markers at either end of a limb segment it is an easy step to calculate the absolute angle of that segment in space. It is not necessary that the two markers be at the extreme ends of the limb segment, so long as they are in line with the long bone axis. Consider Figure 2.22, which shows the outline of a leg with seven anatomical markers in a four-segment three joint system. Markers 1 and 2 define the thigh in the sagittal plane. Note that by convention all angles are measured in a counterclockwise direction starting with horizontal equal to 0°. Thus, θ_{43} is the angle of the shank in space and can be calculated from:

$$\theta_{43} = \arctan \frac{y_3 - y_4}{x_3 - x_4} \qquad (2.13)$$

or, in more general notation:

$$\theta_{ij} = \arctan \frac{y_j - y_i}{x_j - x_i} \qquad (2.14)$$

As has already been noted, these segment angles are absolute in the defined spatial reference system. It is therefore quite easy to calculate the joint angles from the angles of the two adjacent segments.

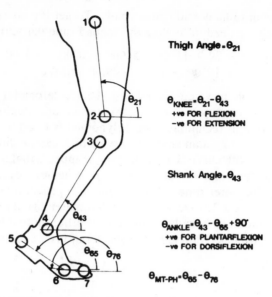

Figure 2.22   Marker locations and limb and joint angle definitions for data in Table 2.3. (By permission of *Journal of Biomechanics*.)

2.6.2   Joint Angles

Each joint has a convention for describing its magnitude and polarity. For example, when the knee is fully extended it is described as 0° flexion, and when the shank moves in a posterior direction relative to the thigh the knee is said to be in flexion. In terms of the absolute angles described previously:

$$\text{knee angle} = \theta_k = \theta_{21} - \theta_{43}$$
$$\text{if } \theta_{21} > \theta_{43} \text{ the knee is flexed}$$
$$\text{if } \theta_{21} < \theta_{43} \text{ the knee is extended}$$

The convention for the ankle is slightly different in that 90° between the shank and foot is the boundary between plantarflexion and dorsiflexion. Therefore,

$$\text{ankle angle} = \theta_a = \theta_{43} - \theta_{65} + 90°$$
$$\text{if } \theta_a \text{ is +ve the foot is plantarflexed}$$
$$\text{if } \theta_a \text{ is -ve the foot is dorsiflexed}$$

2.6.3   Example Problem

For the locomotion data given in Table 2.3 confirm the calculation for the following angles in Field 13.

(i)   Calculate ankle angle of left and right foot.

(ii)   Repeat, assuming marker 6 was not used, and the angle of the foot was defined as that joining the heel to the toe. What error results from this, compared to the method used in part (i)?

Table 2.3   Data For Calculation of Joint Angles, Velocities and Accelerations

	FIELD	MARKER	1	2	3	4	5	6	7
	11	X	110.75	107.46	103.84	76.74	68.05	78.21	88.92
		Y	104.26	68.27	59.99	34.78	32.37	21.73	18.58
	12	X	113.56	111.63	108.02	79.43	70.61	79.30	89.25
		Y	104.09	68.03	59.80	36.68	35.55	23.37	18.66
RIGHT	13	X	116.33	116.02	112.53	82.53	73.73	80.96	90.07
		Y	103.95	67.86	59.73	38.70	38.80	25.33	18.91
LEG	14	X	119.01	120.52	117.23	85.98	77.31	83.31	91.59
(SWING)		Y	103.86	67.86	59.84	40.73	41.86	27.46	19.38
	15	X	121.58	125.03	121.98	89.75	81.23	86.39	94.02
		Y	103.86	68.08	60.17	42.61	44.45	29.53	20.09
	11	X	114.94	129.36	131.81	134.83	129.58	145.20	157.51
		Y	107.46	74.15	65.62	26.99	19.52	19.15	21.02
LEFT	12	X	117.16	131.32	133.50	134.94	129.51	145.17	157.54
		Y	108.06	74.45	65.76	27.01	19.52	19.03	20.68
LEG	13	X	119.32	133.07	134.97	135.02	129.46	145.14	157.55
(STANCE)		Y	108.58	74.64	65.85	27.06	19.53	19.01	20.53
	14	X	121.46	134.61	136.28	135.11	129.47	145.13	157.59
		Y	109.02	74.73	65.89	27.13	19.53	19.04	20.49
	15	X	134.65	135.98	137.47	135.24	129.54	145.14	157.65
		Y	109.37	74.74	65.89	27.19	19.55	19.10	20.51

	RIGHT LEG				LEFT LEG			
ANGLE FIELD	θ_{21}	θ_{43}	θ_{65}	θ_{76}	θ_{21}	θ_{43}	θ_{65}	θ_{76}
11	84.77	42.94	133.68	163.62	113.41	94.46	178.63	188.64
12	86.68	38.96	125.50	154.71	112.85	92.12	178.20	187.58
13	89.52	35.02	118.24	144.81	112.04	90.07	178.11	186.96
14	92.41	31.45	112.62	135.72	110.98	88.27	178.18	186.63
15	95.52	28.58	109.08	128.93	109.71	86.71	178.33	186.45

(iii) When during the gait cycle would you predict that this error is greatest?

Answer

(i)   Right Leg

$$\theta_{43} = \arctan \frac{59.73 - 38.70}{112.53 - 82.53} = 35.03°$$

$$\theta_{65} = \arctan \frac{38.80 - 25.33}{73.73 - 80.96} = 118.22°$$

$\theta_a = \theta_{43} - \theta_{65} + 90° = 35.03° - 118.22° + 90° = 6.81°$
(plantarflexion)

Left Leg

$$\theta_{43} = \arctan \frac{65.85 - 27.06}{134.97 - 135.02} = 90.07°$$

$$\theta_{65} = \arctan \frac{19.53 - 19.01}{129.46 - 145.14} = 178.10°$$

$\theta_a = \theta_{43} - \theta_{65} + 90° = 90.07° - 178.10° + 90° = 1.97°$
(plantarflexion)

(ii) Right Leg

$$\theta_{75} = \arctan \frac{38.8 - 18.91}{73.73 - 90.07} = 129.4°$$

$\theta_a = \theta_{43} - \theta_{75} + 90° = 32.38° - 129.4° + 90° = -7.02°$
(dorsiflexion)

This represents an error of $6.81° + 7.02° = 13.83°$

Left Leg

$$\theta_{75} = \arctan \frac{19.53 - 20.53}{129.46 - 157.55} = 182.04°$$

$\theta_a = \theta_{43} - \theta_{75} + 90° = 90.07° - 182.04° + 90° = -1.97°$
(dorsiflexion)

This represents an error of $1.97° + 1.97° = 3.94°$

(iii) This error would be the greatest at push-off, as the body weight causes the toe to extend about the metatarsophalangeal joints. Thus the toe marker does not represent the distal end of the rigid foot segment.

2.7   CALCULATION OF VELOCITY AND ACCELERATION

As seen in Section 2.5.4, there can be severe problems associated with the determination of velocity and acceleration information. For the reasons outlined we will assume that the raw displacement data has been suitably smoothed by digital filtering and we have a set of smoothed coordinates and angles to operate upon.

To calculate the velocity from displacement data all that is needed is to take the finite difference. For example, to determine the velocity in the x direction we calculate $\Delta x / \Delta t$,

where:

$\Delta x = x_{i+1} - x_i$
Δt is the time between adjacent samples, x_{i+1} and x_i

The velocity calculated this way does not represent the velocity at either of the sample times. Rather, it represents the velocity of a point in time halfway between the two samples. This can result in errors later on when we try to relate the velocity-derived information to displacement data, and both results do not occur at the same point in time. A way around this problem is to calculate the velocity and accelerations on the basis of $2\Delta t$ rather than Δt. Thus the velocity at the ith sample is:

$$Vx_i = \frac{x_{i+1} - x_{i-1}}{2\Delta t} \tag{2.15}$$

Note that the velocity calculated is at a point halfway between the two samples, as depicted in Figure 2.23. The assumption is that the line joining x_{i-1} to x_{i+1} has the same slope as the line drawn tangent to the curve at x_i.

Similarly, the acceleration is:

$$Ax_i = \frac{Vx_{i+1} - Vx_{i-1}}{2\Delta t} \tag{2.16}$$

2.7.1   Example Problem

For the locomotion data given, calculate the velocity and acceleration of the heel of the right leg in the horizontal direction for field 13.

$$Vx_{13} = \frac{x_{14} - x_{12}}{2\Delta t} = \frac{77.31 - 70.61}{2 \times 1/60} = 201 \text{ cm/sec}$$

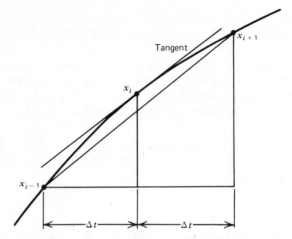

Figure 2.23   Finite difference techniques for calculating slope of curve at the ith sample point.

Before we can calculate the acceleration at frame 13 we must first calculate the velocity at fields 14 and 12.

$$Vx_{14} = \frac{x_{15} - x_{13}}{2\Delta t} = \frac{81.23 - 73.73}{2 \times 1/60} = 225 \text{ cm/sec}$$

$$Vx_{12} = \frac{x_{13} - x_{11}}{2\Delta t} = \frac{73.73 - 68.05}{2 \times 1/60} = 170.4 \text{ cm/sec}$$

$$Ax_{13} = \frac{Vx_{14} - Vx_{12}}{2\Delta t} = \frac{225 - 170.4}{2 \times 1/60} = 1633 \text{ cm/sec}^2$$

2.7.2   Example Problem

For field 13 of the data for the right shank calculate the angular acceleration.

$$\omega_{12} = \frac{\theta_{13} - \theta_{11}}{2\Delta t} = \frac{35.02 - 42.94}{2 \times 1/60} = -237°/\text{sec}$$

$$\omega_{14} = \frac{\theta_{15} - \theta_{13}}{2\Delta t} = \frac{28.58 - 35.02}{2 \times 1/60} = -193°/\text{sec}$$

$$\alpha_{13} = \frac{\omega_{14} - \omega_{12}}{2\Delta t} = \frac{-193 - (-237)}{2 \times 1/60} = 1320°/\text{s}^2 = 23 \text{ rad/sec}^2$$

2.8   PROBLEMS BASED ON KINEMATIC DATA

All the problems in this section utilize live data taken during gait studies in the Gait Laboratory in the Department of Kinesiology at the University of

Waterloo. The raw and smoothed data taken from the right leg and trunk during 1½ strides are presented in Appendix A. Derived kinematic data is also calculated and presented in subsequent tables in the appendix. Complete details for marker location, scaling, and so on, are given in Table A.1.

1. Plot on as large a vertical scale as possible the vertical displacement of the greater trochanter versus time for one-half of a walking stride starting with heel-strike. Plot the raw and filtered vertical coordinates on the same curve. Discuss the results of the digital filtering.

2. (a) Plot the vertical displacement of the heel marker versus time for one complete stride starting with heel-strike. Use filtered data. At maximum heel-rise how high does the heel lift off the ground? Compare this value with the height of the knee above the ground at the same time.

   (b) By scanning the filtered X and Y coordinates of the heel marker, at what frame would you estimate heel-strike?, heel-off?

3. Plot on the same graph the trajectory of the hip and trunk markers for one stride, commencing with heel-strike. Discuss the trajectory plots from the following points:

   (a) Is the shape of these curves what you would expect in walking?

   (b) Is there any evidence of conservation of mechanical energy in these curves (e.g., is potential energy being converted into kinetic energy and vice versa?). Students who are not familiar with mechanical energy analysis should refer to the introductory sections of Chapter 5 before attempting to answer.

4. Plot on the same graph the trajectory of the heel and toe markers for one stride, commencing with heel-strike.

   (a) Note and describe the trajectory of the heel marker in mid-swing. What happens to the trajectory as heel-strike is approached?

   (b) Note the trajectory of the toe marker between heel-strike and flat-foot. Describe the action of the foot during this period. What muscle would you predict is active at this time?

   (c) Determine how much the toe clears the ground at its lowest point during swing. (Hint: The height of the toe marker above ground level can be determined from its average Y coordinate shortly before toe-off.)

5. **(a)** Calculate the maximum velocity of the ankle and toe (by first scanning their X and Y filtered coordinates). Calculate the direction of these velocity vectors.

(b) For the same frame as in **(a)** calculate the horizontal and vertical acceleration of the knee and hip (greater trochanter).

(c) Also, for the same frame calculate the shank angle, angular velocity, and acceleration.

6. Determine the average forward velocity of the body from any appropriate kinematic data.

REFERENCES

Dinn, D. F., Winter, D. A. and Trenholm, B. G. CINTEL-Computer Interface for Television. *IEEE Trans. on Computers* **C-19:**1091–1095, 1970.

Eberhart, H. D. and Inman, V. T. An evaluation of experimental procedures used in a fundamental study of human locomotion. *Ann. N.Y. Acad. Sci.* **5:**1213–1228, 1951.

Finley, F. R. and Karpovich, P. V. Electrogoniometric analysis of normal and pathological gaits. *Res. Quart.* **35:**379–384, 1964.

Morris, J. R. W. Accelerometry—a technique for the measurement of human body movements. *J. Biomech.* **6:**729–736, 1973.

Murray, M. P., Drought, A. B. and Kory, R. C. Walking patterns of normal men. *J. Bone Joint Surg.* **48A:**335–360, 1964.

Öberg, K. Mathematical modeling of human gait: an application of the SELSPOT system. *Biomechanics IV*, R. C. Nelson and C. A. Moorehouse (Eds.). University Park Press, pp. 79–84. Baltimore, 1974.

Pezzack, J. C., Norman, R. W. and Winter, D. A. An assessment of derivative determining techniques used for motion analysis. *J. Biomech.* **10:**377–382, 1977.

Radar, C. M. and Gold, B. Digital filtering design techniques in the frequency domain. *Proc. IEEE*, **55:**149–171, 1967.

Winter, D. A., Greenlaw, R. K. and Hobson, D. A. Television-computer analysis of kinematics of human gait. *Computers Biomed. Res.* **5:**498–504, 1972.

Winter, D. A., Sidwall, H. G. and Hobson, D. A. Measurement and reduction of noise in kinematics of locomotion. *J. Biomech.* **7:**157–159, 1974.

Winter, D. A. and Wells, R. P. Proper sampling and filtering frequencies in the kinematics of human gait. *Proc. 7th Cdn. Med. Biol. Eng. Conf.*, Vancouver, Aug. 1978.

CHAPTER THREE
Anthropometry

3.0 SCOPE OF ANTHROPOMETRY IN MOVEMENT BIOMECHANICS

Anthropometry is the major branch of anthropology that studies the physical development of the human body. A wide variety of physical measurements are required to describe and differentiate the characteristics of race, sex, age, and body type. The major emphasis of these studies in the past has been evolutionary and historical. However, more recently a major impetus has come from the needs of technological developments, especially man-machine interfaces: workspace design, cockpits, pressure suits, armor, and so on. Most of these needs are satisfied by basic linear, area, and volume measures. However, human movement analysis requires kinetic measures as well: masses, moments of inertia, and their locations. There is also a considerable body of knowledge, not all of which is available, regarding the joint centers of rotation, origin and insertion of muscles, angles of pull of tendon, and length and cross-sectional area of muscles.

3.1 SEGMENT DIMENSIONS

The most basic body dimension is the length of segments between each joint. These vary with body build, sex, and racial origin. Dempster (1955, 1959) has summarized estimates of segment lengths and joint center locations relative to anatomical landmarks. An average set of segment lengths expressed as a percent of body height has been prepared by Drillis and Contini (1966) and is shown in Figure 3.1. These segment proportions serve as a good approximation in the absence of better data, preferably measured directly from the individual.

3.2 DENSITY, MASS AND INERTIAL PROPERTIES

Kinematic and kinetic analyses require data regarding mass distributions, mass centers, moments of inertia, and the like. Some of these measures

47

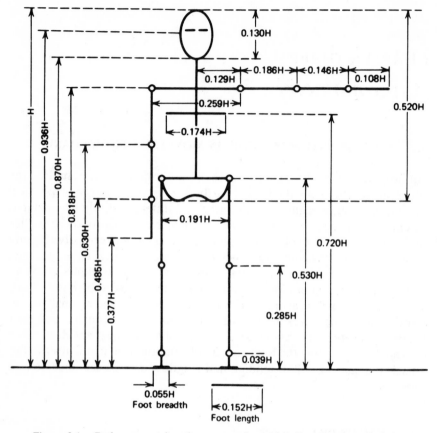

Figure 3.1   Body segment lengths expressed as a fraction of body height, *H*.

have been determined directly from cadavers; others have utilized measured segment volumes in conjunction with density tables.

3.2.1   Whole Body Density

The human body consists of many types of tissue, each with a different density. Cortical bone has a specific gravity over 1.8, muscle tissue is just over 1.0, fat is less than 1.0, and the lungs contain light respiratory gases. The average density is a function of body build, called somatotype. Drillis and Contini (1966) developed an expression for body density, *d,* as a function of ponderal index, $c = h/w$ ⅓, where *w* is body weight (lb) and *h* is body height (in.):

$$d = .69 + .0297c \text{ kg/l} \tag{3.1}$$

The equivalent expression in metric units, where body mass is expressed in kilograms and height in meters, is:

$$d = .69 + 0.9c \text{ kg/l} \tag{3.2}$$

It can be seen that a short fat person has a lower ponderal index than a tall skinny person and therefore has a lower body density.

3.2.2 Segment Densities

Each body segment has a unique combination of bone, muscle, fat, and other tissue, and the density within a given segment is not uniform. Generally, because of the higher proportion of bone, the density of distal segments is greater than proximal segments, and all segments increase their densities as the body density increases. Figure 3.2 shows these trends for six limb segments as a function of whole body density, as calculated by equation 3.1 or 3.2 or as measured directly (Drillis and Contini, 1966; Contini, 1972).

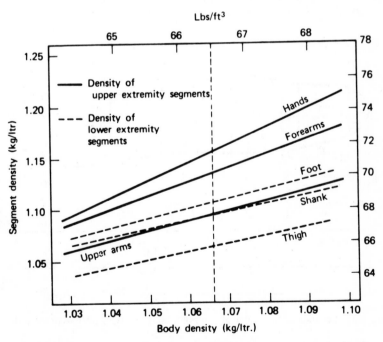

Figure 3.2 Density of limb segments as a function of average body density.

3.2.3   Segment Mass and Center of Mass

As the total body mass increases, so does the mass of each individual segment. Therefore, it is possible to express the mass of each segment as a percent of the total body mass. Table A.2 summarizes the compiled results of several investigators. These values are utilized throughout this text in subsequent kinetic and energy calculations. The location of the center of mass is also given as a percentage of the segment length from either the distal or proximal end. In cadaver studies it is quite simple to locate the center of mass by simply determining the center of balance of each segment. To calculate the center of mass *in vivo* we need the profile of cross-sectional area and length. See Figure 3.3 for a hypothetical profile where the segment is broken into *n* sections, each with its mass indicated. The total mass, *M*, of the segment is:

$$M = \sum_{i=1}^{n} m_i \qquad (3.3)$$

where: m_i is the mass of the ith section
$$m_i = d_i V_i$$

where: d_i is the density of the ith section
V_i is the volume of the ith section

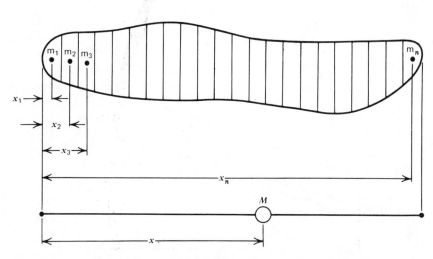

Figure 3.3   Location of the center of mass of a body segment, relative to the distributed mass.

If the density, d, is assumed to be uniform over the segment, then

$$m_i = dV_i$$

and

$$M = d \sum_{i=1}^{n} V_i \qquad (3.4)$$

The center of mass is such that it must have the same net moment of force about any point along the segment axis as did the original distributed mass. Consider the mass center to be located a distance x from the left edge of the segment.

$$Mx = \sum_{i=1}^{n} m_i x_i$$

$$x = \frac{1}{M} \sum_{i=1}^{n} m_i x_i \qquad (3.5)$$

We can now represent the complex distributed mass by a single mass, M, located at a distance x from one end of the segment.

Example 3.1. From the anthropometric data in Table A.2 and the kinematic data in Table A.4 calculate coordinates the center of mass of the thigh and the foot for frame 10.

From Table A.2 the foot center of mass is .5 of the distance from the lateral malleolus (ankle) to the metatarsal marker. From Table A.4, the coordinates of the ankle are (84.9, 11.0) and the metatarsal are (101.1, 1.3)

$$x = (84.9 + 101.1) \div 2 = 93.0 \text{ cm}$$

$$y = (11.0 + 1.3) \div 2 = 6.15 \text{ cm}$$

This center of mass can be verified from Table A.5.

From Table A.2 the thigh center of mass is .433 from the proximal end of the segment. From Table A.4 the coordinates of the proximal end (hip) are (72.1, 92.8) and distal end (knee) are (86.4, 54.9).

$$x = 72.1 + .433 (86.4 - 72.1) = 78.3 \text{ cm}$$

$$y = 92.8 - .433 (92.8 - 54.9) = 76.4 \text{ cm}$$

Again, verification can be made from Table A.7.

3.2.4. Center of Mass of a Multisegment System

With each body segment in motion the center of mass of the total body is continuously changing with time. It is therefore necessary to recalculate

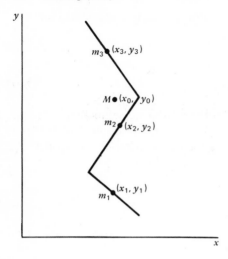

Figure 3.4   Center of mass of a 3-segment system, relative to the centers of mass of the individual segments.

it after each interval of time and this requires a knowledge of the trajectories of the center of mass of each body segment. Consider at a particular point in time a three segment system with the centers of mass as indicated in Figure 3.4. The center of mass of the system is located at (x_0, y_0), and each of these coordinates can be calculated separately:

$$x_0 = \frac{m_1 x_1 + m_2 x_2 + m_3 x_3}{M} \qquad (3.6)$$

$$y_0 = \frac{m_1 y_1 + m_2 y_2 + m_3 y_3}{M} \qquad (3.7)$$

The center of mass (often referred to as center of gravity) of the total body is a frequently calculated variable. Its usefulness in the assessment of human movement, however, is quite limited. Some researchers have used the time history center of mass to calculate the energy changes of the total body. Such a calculation is erroneous, because the center of mass does not reflect reciprocal movements of the limb segments. Thus the energy changes associated with the forward movement of one leg and the backward movement of another may not be detected in the center of mass, which may remain relatively unchanged. More about this will be said in Chapter 5. The major use of the body center of mass is in the analysis of sporting events, especially jumping events where the path of the center of mass is critical to the success of the event.

3.2.5   Mass Moment of Inertia and Radius of Gyration

The location of the center of mass of each segment is needed for an analysis of translational movement through space. If accelerations are involved, we need to know the inertial resistance to such movements. In the linear sense, $F = ma$ describes the relationship between a linear force, F, and the resultant linear acceleration, a. In the rotational sense, $M = I\alpha$. M is the moment of force causing the angular acceleration, α. Thus I is the constant of proportionality that measures the ability of the segment to resist changes in angular velocity. M has units of Newton meters, α is in rads/sec^2 and I is in kg $\cdot$ m^2. The value of I depends on the point about which the rotation is taking place, and is a minimum when the rotation takes place about its center of mass. Consider a distributed mass segment as in Figure 3.3. The moment of inertia about the left end is:

$$I = m_1 x_1^2 + m_2 x_2^2 + \cdot\cdot\cdot\cdot m_n x_n^2$$

$$= \sum_{i=1}^{n} m_i x_i^2$$

It can be seen that the mass close to the center of rotation has very little influence on I while the farthest mass has considerable effect. This principle is used in industry to regulate the speed of rotating machines: the mass of a flywheel is concentrated at the perimeter of the wheel with as large a radius as possible. Its large moment of inertia resists changes in velocity and therefore tends to keep the machine speed constant.

Consider the moment of inertia, I_0, about the center of mass. In Figure 3.5 the mass has been broken into two equal point masses. The location of these two equal components is at a distance ρ from the center such that:

$$I_0 = m\rho^2 \tag{3.9}$$

ρ is called the radius of gyration and is such that the two equal masses

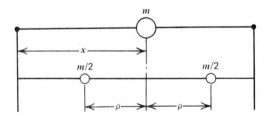

Figure 3.5   Radius of gyration, ρ, of a limb segment, relative to the location of the center of mass of the original system.

shown in Figure 3.5 have the same moment of inertia about the mass center as the original distributed segment did. Note that the center of mass of these two point masses is still the same as the original single mass.

3.2.6 Parallel Axis Theorem

Most body segments do not rotate about their mass centers, but rather about the joint at either end. *In vivo* measures of moment of inertia can only be measured about a joint center; the relationship between this moment of inertia and that about the mass center is given by the parallel axis theorem; a short proof is now given.

$$I = \frac{m}{2} (x-\rho)^2 + \frac{m}{2} (x+\rho)^2$$
$$= m\rho^2 + mx^2$$
$$= I_0 + mx^2 \qquad (3.10)$$

where: I_0 is the moment of inertia about the mass center.

x is the distance between the mass center and the center of rotation.

m is the mass of the segment.

Actually, x can be any distance in either direction from the mass center as long as it lies along the same axis as I_0 was calculated.

Example 3.2. (a) A prosthetic leg has a mass of 3 kg and a center of mass 20 cm from the knee joint. The radius of gyration is 14.1 cm. Calculate I about the knee joint.

$$I_0 = m\rho^2 = 3 \times (.141)^2 = .06 \text{ kg} \cdot \text{m}^2$$
$$I = I_0 + mx^2$$
$$= .06 + 3 \times .2^2 = .18 \text{ kg} \cdot \text{m}^2$$

(b) If the distance between the knee and hip joints is 42 cm calculate the moment of inertia of this prosthesis about the hip joint as the amputee swings through with a locked knee.

x = Distance from mass center to hip = 20 + 42 = 62 cm
$$I = I_0 + mx^2$$
$$= .06 + 3 \times .62^2 = 1.21 \text{ kg} \cdot \text{m}^2$$

3.3 USE OF ANTHROPOMETRIC TABLES AND KINEMATIC DATA

Using Table A.2 in conjunction with kinematic data we can calculate many variables needed for kinetic or energy analyses (Chapters 4 and 5).

Table A.1 gives the basic anthropometric data for the subject whose data is reported. Table A.2, as has been indicated before, gives the segment mass as a fraction of body mass and mass centers as a fraction of their lengths from either the proximal or distal end. The radius of gyration is also expressed as a fraction of the segment length about the center of mass, the proximal end, and the distal end.

3.3.1 Calculation of Segment Masses and Mass Centers

Example 3.3. Calculate the mass of the foot, shank, thigh, and H.A.T., and its location from proximal or distal end.

From Table A.1, the body mass is 80 kg; fractions are obtained from Table A.2.

Mass of foot   = .0145 × 80 = 1.16 kg
Mass of shank   = .0465 × 80 = 3.72 kg
Mass of thigh   = .10 × 80 = 8.0 kg
Mass of H.A.T. = .678 × 80 = 54.24 kg
From Table A1: foot = .195 m, shank = .435 m, thigh = .410 m, H.A.T. = .295 m
C of M of foot   = .50 × .195 = .098 m from ankle marker
C of M of shank = .433 × .435 = .188 m from femoral condyle marker
C of M of thigh   = .433 × .410 = .178 m from greater trochanter marker
C of M of H.A.T. = 1.142 × .295 = .337 m above greater trochanter marker

3.3.2 Calculation of Total Body Center of Mass

The calculation of the center of mass of the total body is a special case of Equations 3.6 and 3.7. For an *n* segment body system the center of mass in the X direction is:

$$x = \frac{m_1 x_1 + m_2 x_2 + \cdots \cdot m_n x_n}{m_1 + m_2 + \cdots m_n} \qquad (3.11)$$

where: $m_1 + m_2 + \cdots m_n = M$, the total body mass.

It is quite normal to know the values of m_1, m_2, and so on, as fractions of the total body mass: $m_1 = f_1 M$, $m_2 = f_2 M$, and so on
Therefore,

$$x = \frac{f_1 M x_1 + f_2 M x_2 + \cdots f_n M x_n}{M} \qquad (3.12)$$

$$= f_1 x_1 + f_2 x_2 + \cdots f_n x_n$$

This equation is easier to use because all we require is a knowledge of the fraction of total body mass and the coordinates of each segment's center of mass. These fractions are given in Table A.2.

It is not always possible to measure the center of mass of every segment, especially if it is not in full view of the camera. In the example data presented we have the kinematics of the right side of H.A.T. and the right leg. It is therefore necessary to simulate data for the left side of H.A.T. and the left leg. If we assume symmetry of gait, we can say that the trajectory of the left leg is the same as the right leg but out of phase by half a stride. Thus if we use data for the right leg one half stride later in time and shift it back in space one half a stride length, we can simulate data for the left leg and left side of H.A.T.

Example 3.4.   Calculate the total body center of mass at frame 15.

The time for one stride = 68 frames (HSR = frame 1, the next HSR = frame 69). Thus the data from frame 15 become the data for the right leg and H.A.T., and the data one half stride (34 frames) later become that for the left leg. All coordinates for the leg from frame 49 must now be shifted back in the x direction by a step length. An examination of the x coordinates of the heel during two successive periods of stance shows the stride length to be 236.6 − 78.0 = 158.6 cm. Therefore the step length = 79.3 cm = .793 m. Tabulated below are the coordinates of the body segments for frame 15 (see Tables A.5–A.8).

Segment	X (meters)		Y (meters)	
	Right	Left	Right	Left
Foot	.929	1.353 − .793 = .56	.062	.156
Shank	.884	1.536 − .793 = .743	.358	.416
Thigh	.863	1.639 − .793 = .846	.773	.760
½ H.A.T.	.791	1.584 − .793 = .791	1.275	1.265

$x = .014(.929+.56) + .047(.884+.743) + .1(.863+.846) + .339(.791+.791)$

$= .805$ (compare with computer calculation of .812, Table A.9)

$y = .014(.062 + .156) + .047(.358 + .416) + .1(.773 + .760) +$
$.339(1.275 + 1.265)$

$= 1.054$ (compare with computer calculation of 1.058, Table A.9)

3.3.3   Calculation of Moment of Inertia

Example 3.5.   Calculate the moment of inertia of the shank about its center of mass, its distal end, and its proximal end.

From Table A.2 the mass of the shank is $.0465 \times 80 = 3.72$ kg. The shank length (Table A.1) is given as .435 m. The radius of gyration/segment length is .302 for the center of mass, .528 for the proximal end, and .643 for the distal end.

$$I_0 = 3.72 \ (.435 \times .302)^2 = .064 \ kg \cdot m^2$$

About proximal end,

$$I = 3.72 \ (.435 \times .528)^2$$
$$= .196 \ kg \cdot m^2$$

About distal end,

$$I = 3.72 \ (.435 \times .643)^2$$
$$= .291 \ kg \cdot m^2$$

Example 3.6. Calculate the moment of inertia of H.A.T. about its proximal end and about its center of mass.

From Table A.2 the mass of H.A.T. is $.678 \times 80 = 54.24$ kg. The H.A.T. length (Table A.1) is given as .295 m. The radius of gyration about the proximal end/segment length is 1.456.

$$I_p = 54.24 \times (.295 \times 1.456)^2$$
$$= 10.00 \ kg \cdot m^2$$

From Table A.2 the center of mass/segment length = 1.142 from the proximal end.

$$I_0 = I_p - mx^2$$
$$= 10.00 - 54.24 \ (.295 \times 1.142)^2$$
$$= 3.84 \ kg \cdot m^2$$

Or we could use the radius of gyration/segment length about the C of G = .903.

$$I_0 = m\rho^2$$
$$= 54.24 \ (.295 \times .903)^2$$
$$= 3.84 \ kg \cdot m^2$$

3.4 DIRECT EXPERIMENTAL MEASURES

For more exact kinematic and kinetic calculations it is preferable to have directly measured anthropometric values. The equipment and techniques that have been developed have limited capability and sometimes are not much of an improvement over the values obtained from tables.

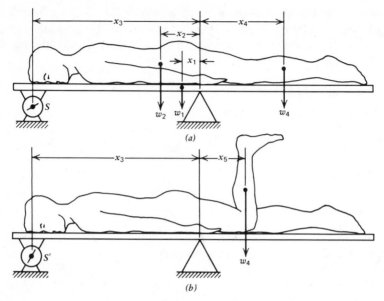

Figure 3.6   Balance board technique for the *in vivo* determination of the mass of a distal segment. See text for details.

3.4.1   Location of Anatomical Center of Mass of the Body

The center of mass of the total body, called the anatomical center of mass, is readily measured using a balance board, as shown in Figure 3.6a. It consists of a rigid board mounted on a scale at one end and a pivot point at the other end, or at some convenient point on the other side of the body's center of mass. There is an advantage in locating the pivot as close as possible to the mass center. A more sensitive scale (0–5 kg) rather than a 50- or 100-kg scale is possible, resulting in greater accuracy. It is presumed that the weight of the balance board, W_1, and its location from the pivot, x_1, are both known along with the body weight, W_2. With the body lying prone the scale reading is S (an upward force acting at a distance x_3 from the pivot). Taking moments about the pivot:

$$W_1 x_1 + W_2 x_2 = S x_3$$

$$x_2 = \frac{S x_3 - W_1 x_1}{W_2} \tag{3.13}$$

3.4.2 Calculation of Mass of a Distal Segment

The mass or weight of a distal segment can be determined by the technique demonstrated in Figure 3.6b. The desired segment, here the leg and foot, is lifted to a vertical position so that its mass center lies over the joint center. Prior to lifting the center of mass was x_4 from the pivot point, with the scale reading S. After lifting, the leg center of mass is x_5 from the pivot, and the scale reading has increased to S^1. The decrease in the clockwise moment due to the leg movement is equal to the increase in the scale reaction force moment about the pivot point:

$$W_4(x_4 - x_5) = (S - S^1)x_3$$
$$W_4 = \frac{(S - S^1)x_3}{(x_4 - x_5)} \tag{3.14}$$

The major error in this calculation is due to errors in x_4, usually obtained from anthropometric tables. To get the mass of the total lower limb this experiment can be repeated with the subject lying on his back with the limb flexed at an angle of 90°. From the mass of the total limb we can now subtract that of the shank and foot to get the thigh mass.

3.4.3 Moment of Inertia of a Distal Segment

The equation for the moment of inertia, described in Section 3.25, can be used to advantage as a means of calculating I at a given joint center of rotation. I is the constant of proportionality that relates the joint moment to the segment's angular acceleration, assuming the proximal segment is fixed. A method called the "quick-release" experiment can be used to calculate I directly, and requires the arrangement pictured in Figure 3.7. We know that $I = M/\alpha$, so if we can measure the moment, M, that causes an angular acceleration, α, we can calculate I directly. A horizontal force, F, pulls on a convenient rope or cable at a distance y_1 from the joint center, and is restrained by an equal and opposite force acting on a release mechanism. An accelerometer is attached to the leg at a distance y_2 from the joint center. The tangential acceleration, a, is related to the angular acceleration of the leg, α, by: $a = y_2\alpha$.

With the forces in balance as shown, the leg is held in a neutral position and no acceleration occurs. If the release mechanism is actuated the restraining force suddenly drops to zero and the net moment acting on the leg is Fy_1, which causes an instantaneous acceleration α. F and a can be recorded on a dual beam storage oscilloscope; most pen recorders have

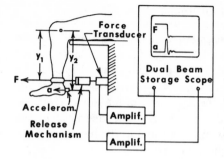

Figure 3.7   Quick-release technique for the determination of the mass moment of inertia of a distal segment. Force, F, applied horizontally results, after release of the segment, in an initial acceleration, a. Moment of inertia can then be calculated from F, a, y_1 and y_2.

too low a frequency response to capture the acceleration impulse. The moment of inertia can now be calculated:

$$I = \frac{M}{\alpha}$$

$$= \frac{Fy_1 y_2}{a} \tag{3.15}$$

Figure 3.7 shows the sudden burst of acceleration accompanied by a rapid decrease in the applied force F. This force drops after the peak of acceleration and does so because the forward displacement of the limb causes the tension to drop in the pulling cable. A convenient release mechanism can be achieved by suddenly cutting the cable or rope that holds back the leg. The sudden accelerometer burst can also be used to trigger the oscilloscope sweep so that the rapidly changing force and acceleration can be captured.

More sophisticated experiments have been devised to measure more than one parameter simultaneously. Such techniques have been developed by Hatze (1975) and are capable of determining the moment of inertia, location of mass center, and damping coefficient simultaneously.

3.5 MUSCLE ANTHROPOMETRY

Before we can calculate the forces produced by individual muscles during normal movement we usually need some dimensions from the muscles themselves. If muscles of the same group share the load they probably do so proportionally to their relative cross-sectional areas. Also, the mechanical advantage of each muscle can be different depending on the moment arm length at its origin and insertion.

Table 3.1   Mass, Length, and Cross-sectional Area of Some Muscles

Muscle	Mass (gm)	Length (cm)	Cross-Section Area (cm²)
Sartorius	113	45	2.4
Semitendinosus	133	15	8.5
Biceps femoris (short head)	65	12	5.2
Soleus	264		67
Gastrocs (Med. Hd.)	142		22
Gastrocs (Lat. Hd.)	91		13
Tibialis posterior	62		17
Tibialis anterior	110		14
Rectus femoris	178		30
Vastus lateralis	410		43
Vastus medialis	324		34
Vastus intermedius	254		28

3.5.1   Cross-Sectional Area and Length of Muscles

A variety of sources have examined the length and cross-sectional areas of human skeletal muscle. A complete literature search is not possible; however, representative dimensions are tabulated for some of the major muscles of the body in Table 3.1.

3.5.2   Change in Muscle Length During Movement

A few studies have investigated the changes in the length of muscles as a function of the angles of the joints they cross. Grieve and colleagues (1978) in a study on eight cadavers reported percentage length changes of the gastrocs as a function of the knee and ankle angle. The resting length of the gastrocs was assumed to be when the knee was flexed 90° and the ankle was in an intermediate position, neither plantarflexed nor dorsiflexed. With 40° plantarflexion the muscle shortened 8.5% and linearly changed its length to a 4% increase at 20° dorsiflexion. An almost linear curve described the changes at the knee: 6.5% increase at full extension to a 3% decrease at 150° flexion.

3.5.3   Force Per Unit Cross-Sectional Area (Stress)

A wide range of stress values for skeletal muscles has been reported (Haxton, 1944; Alexander and Vernon, 1975). Most of these stress values were measured during isometric conditions and range from 20 N/cm^2 to 100 N/cm^2. These higher values were recorded in pennate muscles, which are those whose fibers lie at an angle from the main axis of the muscle. Such an orientation effectively increases the cross-sectional area above that measured and used in the stress calculation. Haxton (1944) related force to stress in two pennate muscles (gastrocs and soleus) and found stresses as high as 38 N/cm^2. Dynamic stresses have been calculated in the quadriceps during running and jumping to be about 40 N/cm^2.

3.5.4   Multi-Joint Muscles

A large number of the muscles in the human body pass over more than one joint. In the lower limbs the hamstrings are extensors of the hip and flexors of the knee, the rectus femorus is a combined hip flexor and knee extensor, and the gastrocnemius are knee flexors and ankle plantarflexors. The fiber length of many of these muscles may be insufficient to allow complete movement of both the joints involved (see Section 6.2.1). Elftman (1966) has suggested that many normal movements require lengthening at one joint simultaneously with shortening at the other. Consider the action of the rectus femorus, for example, during early swing in running. This muscle shortens as the hip flexes and lengthens at the knee as the shank swings backward in preparation for swing through. The tension in the rectus femoris simultaneously creates a flexor hip moment (positive work) and an extensor knee moment to decelerate the swinging shank (negative work) and start accelerating it forward. In this way the net change in muscle length is reduced compared with two equivalent single joint muscles, and excessive positive and negative work can be reduced.

3.5.5   Mechanical Advantage of Muscle

The origin of insertion of each muscle defines the angle of pull of the tendon on the bone and therefore the mechanical leverage it has at the joint center. Each muscle has its unique moment arm length which is the length of a line normal to the muscle passing through the joint center. This moment arm length changes with the joint angle. One of the few studies done in this area (Smidt, 1973) reports the average moment arm length (26 subjects) for the knee extensors and for the hamstrings acting at the knee.

Both these muscle groups showed an increase in the moment length as the knee was flexed, reaching a peak at 45°, then decreasing again as flexion increased to 90°. Wilkie (1950) has also documented the moment arm lengths for elbow flexors.

3.6   STUDENT PROBLEMS

1. (a) Calculate your own average body density.

   (b) Based on your body density in part (a), calculate your body volume in liters.

2. From Figure 3.2 determine the density of your six limb segments.

3. The following circumference measurements were taken on a thigh segment which measured 44 cm in length; eleven measurements were taken, one in the middle of each 4 cm section. Starting at the distal end the circumference in cm was: 46.7, 47.5, 48.2, 49.1, 50.5, 51.8, 53.5, 55.0, 55.8, 56.2, 56.6. Assuming a density of 1.08 gm/cm³ and that the thigh is perfectly circular in cross-section, calculate the location of the center of mass from the distal end.

4. Using the data from question 3, calculate the moment of inertia of the thigh from the distal end. Calculate the radius of gyration about the center of mass.

5. For the subject whose data appears in the appendix calculate the moment of inertia of the following:

   (a) The foot and shank about the knee joint.

   (b) The foot, shank, and thigh about the hip joint.

   (c) The total body as he stands erect about the ankle joint.

   (d) The foot, shank, and thigh, plus a ski boot (whose mass is 2.8 kg and whose mass center is located at the ankle joint) about the hip joint.

6. Calculate the total body center of mass for frame 25 using appropriate data from Appendix A.

7. (a) Assuming the maximal stress generated by the knee extensor muscles to be 40 N/cm², calculate the maximum force generated in the patellar ligament.

   (b) At 45° knee flexion the moment arm length of the patellar tendon about the knee center is 4.8 cm. Calculate the maximum extensor knee moment in N·m.

(c) At full extension the maximum muscle tension drops to 40% of that generated at 45° (because of muscle lengthening beyond its resting length—see Chapter 6). The patellar moment arm length decreases to 4.2 cm. Calculate the maximum knee extensor moment that can be generated at full extension.

REFERENCES

Alexander, R. McN. and Vernon, A. The dimensions of knee and ankle muscles and the forces they exert. *J. Human Movement Studies* **1:**115–123, 1975.

Contini, R. Body segment parameters, Part II. *Artificial Limbs* **16:**1–19, 1972.

Dempster, W. T. Space requirements of the seated operator. Wright Patterson Air Force Base, WADC-TR-55-159 (1955).

Dempster, W. T., Gabel, W. C. and Felts, W. J. L. The anthropometry of manual work space for the seated subject. *Amer. J. Phys. Anthrop.* **17:**289–317, 1959.

Drillis, R. and Contini, R. Body segment parameters. Report No. 1163-03, Office of Vocational Rehabilitation, Department Health, Education and Welfare, New York, 1966.

Elftman, H. Biomechanics of muscle, with particular application to studies of gait. *J. Bone Joint Surg.* **48-A:**363–377, 1966.

Grieve, D. W., Cavanagh, P. R. and Pheasant, S. Prediction of gastrocnemius length from knee and ankle joint posture. *Proc. VI Internat. Congr. Biomech.* Copenhagen, July 1977.

Haxton, H. A. Absolute muscle force in the ankle flexors of man. *J. Physiol.* **103:**267–273, 1944.

Hatze, H. A new method for the simultaneous measurement of the moment of inertia, the damping coefficient and the location of the centre of mass of a body segment in situ. *Europ. J. Appl. Physiol.* **34:**217–226, 1975.

Smidt, G. L. Biomechanical analysis of knee flexion and extension, *J. Biomech.* **6:**79–92, 1973.

Wilkie, D. R. The relation between force and velocity in human muscle. *J. Physiol.* **110:**249–280, 1950.

CHAPTER FOUR
Kinetics

4.0 ANALYTICAL APPROACH

Chapter 2 has dealt at length with the description of the movement itself, without regard to the forces that caused the movement. The study of these forces is called kinetics; knowledge of the patterns of these forces is necessary for a deeper understanding of the cause of any movement.

Transducers have been developed that can be implanted surgically to measure the force exerted by a muscle at the tendon. However, such techniques have applications only in animal experiments, and even then to a limited extent. It therefore remains that we attempt to calculate these forces indirectly, using readily available kinematic data. The process by which the reaction forces and muscle moments are calculated requires a biomechanical link-segment model. Such a process is depicted in Figure 4.1. If we have a full kinematic description, accurate anthropometric measures, and the external forces, we can calculate the joint reaction forces and muscle moments. This prediction is a very powerful tool in gaining insight into the net pattern of muscle activity. Such information is very useful to the coach, therapist, and kinesiologist in their diagnostic assessments. The effect of training, therapy, or surgery is extremely evident at this level of assessment, although it is often obscured in the original kinematics.

4.1 THE LINK SEGMENT MODEL

The accuracy of any assessment is only as good as the model itself. Accurate measures of segment masses, centers of mass, joint centers, and moments of inertia are required. Such data can be obtained from statistical tables based on the person's height, weight, and, sometimes, sex, as was detailed in Chapter 3. A limited number of these variables can be measured directly, but some of the techniques are time consuming and have limited accuracy. Regardless of the source of the anthropometric data, the following assumptions are made with respect to the model:

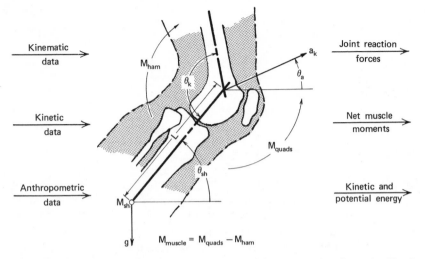

Figure 4.1 Schematic diagram to show the relationship between the kinematic, kinetic, and anthropometric data and the calculated forces, moments, and energies.

1. Each segment has a fixed mass located as a point mass at its center of gravity.

2. The location of the center of gravity remains fixed during the movement.

3. The joints are considered to be hinge (pin) joints.

4. The mass moment of inertia of each segment about its mass center (or about either proximal or distal joints) is constant during the movement.

Figure 4.2 shows the equivalence between the anatomical and link-segment models for a leg. The segment masses, m_1, m_2 and m_3 are considered to be concentrated at points. The distance from the proximal joint to the mass centers is considered to be fixed, as are the length of the segments.

4.1.1   Forces Acting on the Link Segment Model

1. Gravitational Forces. The forces of gravity act downward through the centers of mass of each segment and are equal to the magnitude of the mass × acceleration due to gravity (normally 9.8 m/sec²).

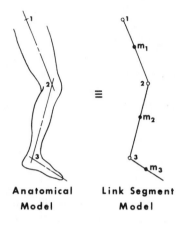

Figure 4.2   Relationship between the anatomical model and the link-segment model. Joints are replaced by hinge (pin) joints and segments are replaced by masses located at mass centers.

Anatomical Model   **Link Segment Model**

2.   Ground Reaction or External Forces.   Any external forces must be measured by an external force transducer. Such forces are distributed over an area of the body (such as the ground reaction forces under the area of the foot). In order to represent such forces as vectors they must be considered to act at a point, which is usually called the center of pressure. A suitably constructed force plate, for example, yields signals from which the center of pressure can be calculated.

3.   Muscle Forces.   The net effect of muscle activity at a joint can be calculated in terms of net muscle moments. If a co-contraction is taking place at a given joint, the analysis yields only the net effect of both agonist and antagonistic muscles. Also, any friction effects of the joints or within the muscle cannot be separated from this net value. Increased friction merely reduces the effective "muscle" moment; the muscle contractile elements themselves are actually creating moments higher than that analyzed at the tendon. However, the error at low and moderate speeds of movement is usually less than 5%.

4.1.2   Joint Reaction Forces and Bone-on-Bone Forces

The three forces described above constitute all the forces acting on the total body system itself. However, our analysis examines the segments one at a time and therefore must calculate the reaction forces between segments. A free body diagram of each segment is required, as shown in Figure 4.3. Here the original link segment model is broken into its segmental parts. For convenience we make the break at the joints, and the forces that act across each joint must be shown on the resultant free body diagram. This procedure now permits us to look at each segment and

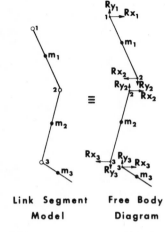

Link Segment   Free Body
Model   Diagram

Figure 4.3 Relationship between free body diagram and the link-segment model. Each segment is "broken" at the joints and the reaction forces at the joints are represented as indicated.

calculate all unknown joint reaction forces. In accordance with Newton's third law, there is an equal and opposite force acting at each hinge joint in our model. For example, when a leg is held off the ground in a static condition, the foot is exerting a downward force on the tendons and ligaments crossing the ankle joint. This is seen as a downward force acting on the shank equal to the weight of the foot. Likewise, the shank is exerting an equal upward force on the foot through the same connective tissue.

Considerable confusion exists regarding the relationship between joint reaction forces and joint bone-on-bone forces. The latter forces are the actual forces seen across the articulating surfaces and include the effect of muscle activity. Actively contracting muscles pull the articulating surfaces together, creating a compressive force. Thus bone-on-bone forces = active force due to muscle + joint reaction forces. Figure 4.4 illustrates these differences in a simple situation. Case 1 has the lower segment with a weight of 100 N hanging passively from the muscles originating in the upper segment. The two muscles are not contracting but are, assisted by ligamentous tissue, pulling upward with an equal and opposite 100 N. The link segment model shows these equal and opposite reaction forces. The bone-on-bone force is zero, indicating that the joint articulating surfaces are under neither tension nor compression. In Case 2 there is an active contraction of the muscles, so that the total upward force is now 170 N. The bone-on-bone force is 70 N compression. This means that a force of 70 N exists across the articulating surfaces. As far as the lower segment is concerned, there is still a net reaction force of 100 N upward (170 N upward through muscles and 70 N downward across the articulating surfaces). The lower segment is still acting downward with a force of 100

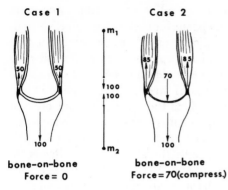

Figure 4.4 Diagrams to illustrate the difference between joint reaction forces and bone-on-bone forces. In both cases, as shown, the reaction force is 100 N acting upward on M_2 and downward on M_1. With no muscle activity, the bone-on-bone force is zero; with muscle activity, as shown in case 2, the bone-on-bone force is 70 N.

N; thus the free body diagram remains the same. Generally, the anatomy is not as simple as depicted. More than one muscle is usually active on each side of the joint, so it is difficult to apportion the forces among the muscles. Also, the angle of pull of each tendon and the geometry of the articulating surfaces is not always readily available.

There are common situations when muscles act on one side of a joint to stabilize it against gravity, and it is at this time that the bone-on-bone forces can be several times body weight. In Figure 4.5 we see the knee

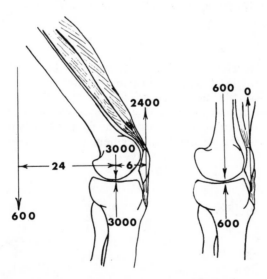

Figure 4.5 Influence of joint angle and muscle forces on the knee bone-on-bone forces during static single support. With no knee flexion, the quadriceps are inactive and the bone-on-bone forces = joint reaction forces = 600 N. With quadriceps tension during knee flexion, the bone-on-bone forces are shown to increase to 5 times the previous value.

joint supporting body weight (600 N) in two different positions: first, with the body center of gravity acting 24 cm posterior to the knee joint; second, with body weight passing through the knee center. In the latter situation there is no need for muscle activity (quadriceps) to stabilize the knee, so the tension in the quads is zero. Thus the bone-on-bone force equals the reaction force, which is 600 N (compression). In the former situation there is tension in the quads in order to stabilize the knee joint and prevent collapse. Assuming that the patellar ligament is 6 cm anterior to the knee joint center, it is easy to calculate that the tension in this tendon is 2400 N. This 2400 N, which acting vertically, pulls the articulating surfaces together so that the compressive forces are now 2400 + 600, or 3000 N. This is five times the body weight. Such high forces are commonplace and must be considered as a cause of sporting injuries and as a source of pain and damage in joint pathologies (arthritis and joint replacements).

4.2 BASIC LINK SEGMENT EQUATIONS—THE FREE BODY DIAGRAM
(Representative paper—Bresler and Frankel, 1950)

Each body segment acts independently under the influence of reaction forces and muscle moments acting at either end, plus the forces due to gravity. Consider the planar movement of a segment in which the kinematics, anthropometrics, and reaction forces at the distal end are known. See Figure 4.6.

Known

a_x, a_y —acceleration of the segment C of G
θ —angle of segment in plane of movement
α —angular acceleration of segment in plane of movement
R_{xd}, R_{yd}—reaction forces acting at distal end of segment, usually determined from a prior analysis of the proximal forces acting on the distal segment

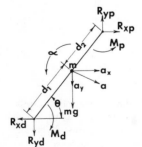

Figure 4.6 A complete free body diagram of a single segment, showing reaction and gravitational forces, net moments of force, and all linear and angular accelerations.

M_d —net muscle moment acting at the distal joint, usually determined from an analysis of the proximal muscle acting on the distal segment.

Unknown

R_{xp}, R_{yp}—reaction forces acting at the proximal joint
M_p—net muscle moment acting on the segment at the proximal joint.

Equations

1.
$$\Sigma Fx = ma_x \qquad (4.1)$$
$$R_{xp} - R_{xd} = ma_x$$

2.
$$\Sigma Fy = ma_y \qquad (4.2)$$
$$R_{yp} - R_{yd} - mg = ma_y$$

3.   About segment C of G, $\Sigma M = I\alpha$ \qquad (4.3)

Note that the muscle moment at the proximal end cannot be calculated until the proximal reaction forces, R_{xp} and R_{yp}, have first been calculated.

Example 4.1 (Figure 4.7).   In a static situation, a person is standing on one foot on a force plate. The ground reaction force is found to act 4 cm anterior to the ankle joint. Note that convention has the ground reaction force, Ry_1, always acting upward. We also show the horizontal reaction force, Rx_1, to be acting in the positive direction (to the right). If this force actually acts to the left we recognize this by substituting a negative number. The subject's mass is 60 kg, and the mass of the foot is 0.9 kg. Calculate the joint reaction forces and net muscle moment at the ankle.

$$Ry_1 = \text{body weight} = 60 \times 9.8 = 588 \text{ N}$$

1. $\Sigma Fx = ma_x$
$$Rx_2 + Rx_1 = ma_x$$
$$Rx_2 + 0 = 0$$
$$Rx_2 = 0$$

Note that this is a redundant calculation in static conditions.

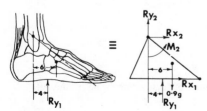

Figure 4.7 Anatomical and free body diagram of foot during weight bearing.

2.
$$\Sigma Fy = ma_y$$
$$Ry_2 + Ry_1 - mg = ma_y$$
$$Ry_2 + 588 - 0.9 \times 9.8 = 0$$
$$Ry_2 = -579.2 \text{ N}$$

The negative sign means that the force acting on the foot at the ankle joint acts *downward*. This is not surprising because the entire body weight, less that of the foot, must be acting downward on the ankle joint.

3. About the C of G of foot, $\Sigma M = I\alpha$
$$M_2 - Ry_1 \times .02 - Ry_2 \times .06 = 0$$
$$M_2 = 588 \times .02 + (-579.2 \times .06)$$
$$= -22.99 \text{ N.m}$$

The negative sign means that the muscle moment acting on the foot at the ankle joint is clockwise. This means that the triceps surae are active via the Achilles tendon of the ankle joint to maintain the static position that resulted in the ground reaction force that was measured.

Example 4.2 (Figure 4.8). From the data in Table A.5 for frame 50 (swing phase), calculate the muscle moment and reaction forces at the ankle.

$$m = 0.0145 \times 80 = 1.16 \text{ kg}$$
$$I = (.0927)^2 \times 1.16 = .00995 \text{ kg} \cdot \text{m}^2$$
$$\alpha = 21.63 \text{ rad/sec}^2$$

1. $\Sigma Fx = ma_x$
$$Rx_1 = 1.16 \times 9.07 = 10.52 \text{ N}$$

2. $\Sigma Fy = ma_y$
$$Ry_1 - 1.16g = m(-6.62)$$
$$Ry_1 = 1.16 \times 9.8 - 1.16 \times 6.62 = 3.69 \text{ N}$$

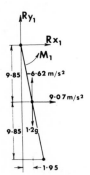

Figure 4.8 Free body diagram of foot during swing (frame 50).

3. At foot C of G $\Sigma M = I\alpha$

$M_1 - Rx_1 \times .0985 - Ry_1 \times .0195 = .00995 \times 21.63$
$M_1 = .00995 \times 21.63 + 10.52 \times .0985 + 3.69 \times .0195$
$\quad = .22 + 1.04 + .07 = 1.33 \text{ N.m}$

Discussion:

(i) The horizontal reaction force of about 11 N is the cause of the horizontal acceleration that we calculated for the foot.

(ii) The foot is decelerating its upward rise at the end of lift-off; thus the vertical reaction force at the ankle is somewhat less than the static gravitational force.

(iii) The ankle muscle moment is positive, indicating net dorsiflexor activity (tibialis anterior), and most of this torque (1.08 out of 1.37 N.m) is required to cause the horizontal acceleration of the foot center of gravity.

Example 4.3 (Figure 4.9). For frame 50 calculate the muscle moments and reaction forces at the knee joint.

$$m = .0465 \times 80 = 3.72 \text{ kg}$$
$$I = (.131)^2 \times 3.72 = .0642 \text{ kg} \cdot \text{m}^2$$
$$\alpha = 36.9 \text{ r/s}^2$$

From Example 4.2,

$$Rx_1 = 10.52 \text{ N}$$
$$Ry_1 = 3.69 \text{ N}$$
$$M_1 = 1.33 \text{ N.m}$$

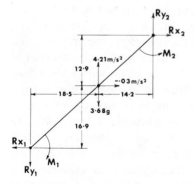

Figure 4.9 Free body diagram of leg during swing (frame 50).

1.
$$\Sigma Fx = ma_x$$
$$Rx_2 - Rx_1 = ma_x$$
$$Rx_2 = 10.52 + 3.72 (-.03) = 10.41 \text{ N}$$

2.
$$\Sigma Fy = ma_y$$
$$Ry_2 - Ry_1 - mg = ma_y$$
$$Ry_2 = 3.69 + 3.72 \times 9.8 + 3.72 (-4.21) = 24.48 \text{ N}$$

3. About shank C of G, $\Sigma M = I\alpha$

$$M_2 - M_1 - .169 \, Rx_1 + .185Ry_1 - .129 \, Rx_2 + .142 \, Ry_2 = I\alpha$$
$$M_2 = 1.33 + .169 \times 10.52 - .185 \times 3.69 + .129 \times 10.41$$
$$- .142 \times 24.48 + .0642 \times 36.9$$
$$= 1.33 + 1.78 - 0.68 + 1.34 - 3.48 + 2.37$$
$$= 2.66 \text{ N.m}$$

Discussion:

(i) M_2 is positive: this represents a counterclockwise (extensor) moment acting at the knee. The quads at this time are rapidly extending the swinging leg.

(ii) The angular acceleration of the shank is the net result of two reaction forces and one muscle acting at each end of the segment. Thus there may not be a single primary force causing the movement we observe; in this case each force had a significant influence on the final acceleration.

4.3   FORCE TRANSDUCERS AND FORCE PLATES

In order to measure the force exerted by the body on an external body or load we need a suitable force-measuring device. Such a device, called a force transducer, gives an electrical signal proportional to the applied force. There are many kinds available: strain gauge, piezoelectric, piezoresistive, capacitive, and others. All these work on the principle that the applied force causes a certain amount of strain in the transducer. For the strain gauge type a calibrated metal plate or beam within the transducer undergoes a very small change (strain) in one of its dimensions. This mechanical deflection, usually less than 1%, causes a change in resistances connected as a bridge circuit (see Section 2.2.2), resulting in an unbalance of voltages proportional to the force. Piezoelectric and piezoresistive types require minute deformations of the atomic structure within a block of special crystalline material. Quartz, for example, is a naturally found piezoelectrical material, and deformation of its crystalline

line structure changes the electrical characteristics of the block of material. The electrical charge across appropriate surfaces of the block is altered, and can be translated via suitable electronics to a signal proportional to the applied force. Piezoresistive types exhibit a change in resistance which, like the strain gauge, upset the balance of a bridge circuit.

4.3.1   Multidirectional Force Transducers

In order to measure forces in two or more directions it is necessary to use a bi- or tridirectional force transducer. Such a device is nothing more than two or more force transducers mounted at right angles to each other. The major problem is to ensure that the applied force acts through the central axis of each of the individual transducers.

4.3.2   Force Plates
(Representative paper—Elftman, 1939)

The most common force acting on the body is the ground reaction force which acts on the foot during standing, walking or running. This force vector is three dimensional, consisting of a vertical component plus two shear components acting along the force plate surface. These shear forces are usually resolved into anterior-posterior and medial-lateral directions.

The fourth variable needed is the location of the center of pressure of this ground reaction vector. The foot is supported over varying amounts of foot surface area with different pressures at each part. Even if we knew these individual pressures over each part of the foot we would be faced with the expensive problem of calculating the net effect as it changes with time. Some attempts have been made to develop suitable pressure-measuring "shoes," but they have been expensive and not too accurate. It is therefore necessary for the force plate to give us this information.

Two common types of force plate are now explained. The first is a flat plate supported by four triaxial transducers, depicted in Figure 4.10. Consider the coordinates of each transducer to be $(_{0,0})$, $(_{0,z})$, $(x_{,0})$ and (x,z). The location of the center of pressure is determined by the relative vertical forces seen at each of these corner transducers. If we designate the vertical forces as F_{00}, Fx_0, F_0z, and Fxz, the total vertical force, $Fy = F_{00} + Fx_0 + F_0z + Fxz$. If all four forces are equal, the center of pressure is at the exact center of the force plate, at $(X/2, Z/2)$. In general,

$$x = \frac{X}{2}\left[1 + \frac{(Fx_0 + Fxz) - (F_{00} + F_0z)}{Fy}\right] \tag{4.4}$$

$$z = \frac{Z}{2}\left[1 + \frac{(F_0z + Fxz) - (F_{00} + Fx_0)}{Fy}\right] \tag{4.5}$$

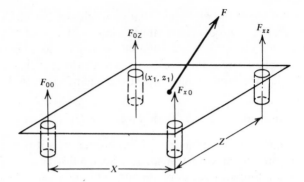

Figure 4.10 Force plate with support in four corners. Magnitude and location of ground reaction force, F, can be determined from the signals from the load cells in each of the support bases.

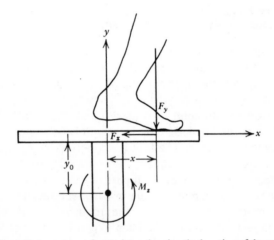

Figure 4.11 Central support type force plate, showing the location of the center of pressure of the foot and the forces and moments involved.

A second type of force plate has one centrally instrumented pillar which supports an upper flat plate. Figure 4.11 shows the forces that act on this instrumented support. The action force of the foot, Fy, acts downward, and the anterior-posterior shear force can act either forward or backward. Consider a reverse shear force, Fx, as shown. If we sum the moments acting about the central axis of the support we get:

$$Mz - Fy \cdot x - Fx \cdot Y_0 = 0$$

$$x = \frac{Mz - Fx \cdot Y_0}{Fy} \qquad (4.6)$$

where: Mz is the bending moment about the axis of rotation of the support.

Y_0 is the distance from support axis to the force plate surface.

Since Fx, Fy, and Mz continuously change with time, x can be calculated to show how the center of pressure moves across the force plate.

Typical force plate data are shown in Figure 4.12 plotted against time for a subject walking at a normal speed. The vertical reaction force, Fy, is very characteristic in that it shows a rapid rise at heel contact to a value in excess of body weight as full weight bearing takes place. Then as the knee flexes during midstance, the plate is partially "unloaded" and Fy drops below body weight. At push-off the leg extends, causing a second peak greater than body weight, and finally the weight drops to zero as the opposite leg takes up the body weight. The anterior-posterior reaction force, Fx, is the horizontal force exerted by the force plate on the foot. Immediately after heel contact it is negative, indicating a backward hori-

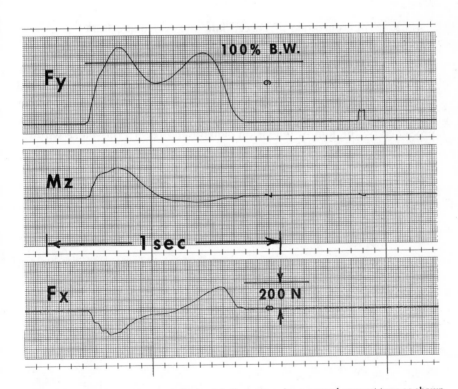

Figure 4.12 Force plate record obtained during gait, using a central support type as shown in Figure 4.11.

zontal friction force between the floor and the shoe. If this force were not present the foot would slide forward; walking on icy or slippery surfaces illustrates this. Near midstance Fx goes positive, indicating that the force plate reaction is acting forward as the muscle forces cause the foot to push back against the plate.

The center of pressure starts at the heel, assuming that initial contact is made by the heel, and then progresses forward toward the ball and toe. The position of the center of pressure relative to the foot cannot be obtained from the force plate data itself; we must first know where the foot is, relative to the axis of the plate. Table A.10 shows typical force plate data as digitized for the sample stride. The type of force plate used to collect these data is the centrally instrumented pillar type depicted in Figure 4.11. The student should note that Mz is positive at heel contact and then becomes negative as the body weight moves forward. The center of pressure, Xcp, and Ycp are calculated in absolute coordinates to match that given in the kinematics listing. Ycp was set at 0 to indicate ground level.

4.3.3 Synchronization of Force Plate and Kinematic Data

Because the kinematic data is usually recorded on film or television it is necessary to synchronize the image data with the time base of the force plate data. This is not always an easy task because film data are recorded on celluloid while force plate data comes as an electrical signal to a pen or tape recorder. The most suitable way is to have a synchronized pulse produced by the film or television camera and record this pulse with the force plate signals.

As well as synchronizing the two time scales it is necessary to find the location of the foot relative to the center of the force plate. Thus the x coordinate of a foot marker is recorded relative to the force plate center.

4.3.4 Combined Force Plate and Kinematic Data

It is valuable to see how the reaction force data from the force plate is combined with the segment kinematics to calculate the muscle moments and reaction forces at the ankle joint during dynamic stance. This is best illustrated in an example calculation for the conditions of frame 35.

Example 4.4 (Figure 4.13)

From equation 4.1, $F_{ax} + F_x = ma_x$
$$F_{ax} = 1.16 \times 3.25 - 160.25$$
$$= -156.5 \text{ N}$$

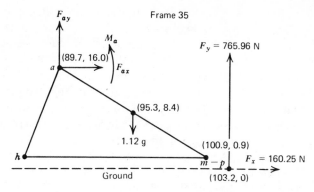

Figure 4.13 Free body diagram of the foot during weight bearing (frame 35).

From equation 4.2, $F_{ay} + F_y - mg = ma_y$
$$F_{ay} = 1.16 \times 1.78 - 765.96 + 1.16 \times 9.81$$
$$= -752.5 \text{ N}$$

From equation 4.3, about the foot C of G

$$\Sigma M = I_0 \alpha$$

$$M_a + F_x \times .084 + F_y \times .079 - F_{ay} \times .056 - F_{ax} \times .076$$
$$= .0099 \, (-45.35)$$

$$M_a = -.0099 \times 45.35 - .084 \times 160.25 - .079 \times 765.96$$
$$- .056 \times 752.5 - .076 \times 156.5$$
$$= -128.44 \text{ N.m.}$$

The polarity and magnitude of this ankle moment indicates fairly strong plantarflexor activity acting to push off the foot and cause it to rotate about the metatarsophalangeal joint.

4.3.5 Interpretation of Muscle Moment Curves
(Representative paper—Pedotti, 1977)

A complete link segment analysis yields the net muscle moment at every joint during the time course of the movement. As an example, we will discuss in detail the ankle moment during one gait stride. Positive moments across the ankle act to dorsiflex the foot; negative moments indicate net plantarflexion. The student is referred to Table B.1 for the moments acting at the right ankle. Slightly before heel strike we see the start of a dorsiflexor (+ve) moment. This creates a stabilizing moment at the ankle in anticipation of heel contact. At heel contact the ground reaction force acts posterior to the ankle joint, and were it not for the dorsiflexor moment the foot would slap down on the floor. Instead, a controlled lowering (negative work) of the foot occurs. Shortly after flat-foot (frame 10) the moment goes negative, meaning that plantarflexor activity is dominant. While the foot is flat on the ground this same muscle activity by the triceps surae decelerates the forward rotation of the shank, preventing excessive breaking of the knee joint. Near midstance the reaction force vector has moved forward of the ankle joint towards the metatarsophalangeal (m−p) joint. If the plantarflexor moment (−ve) is sufficiently large, lifting of the foot about the m−p joint takes place and the ankle moves upward and forward, taking the shank with it. Heel off occurs about frame 20. The knee joint then breaks rapidly as push-off continues, reaching a peak at about frame 33 when the ankle moment is 137 N.m (plantarflexion). At the end of push-off the foot leaves the ground and a dorsiflexor (+ve) moment causes the foot to be lifted so that the toe will clear the ground about mid-swing. The student is encouraged to look at the time histories of the muscle moments at the knee and hip to see if they make sense.

4.4 CALCULATION OF BONE-ON-BONE FORCES DURING DYNAMIC CONDITIONS
(Representative paper—Paul, 1966)

The difference between joint reaction forces and bone-on-bone forces has been dealt with in detail in Section 4.1.2, using a static situation as an example. Dynamic conditions are somewhat more difficult to analyze, and depend on certain assumptions regarding the amount of co-contraction taking place. The net muscle moment that is calculated from the link segment mechanics is the algebraic sum of agonist and antagonist activity. These individual contributions are unknown; however, the following example will illustrate its influence on the bone-on-bone forces.

Example 4.5.   At frame 10 calculate the bone-on-bone forces at the knee joint, assuming the following muscle activities:

(i) The quadriceps only are active, and they produce a vertical force that acts 6 cm from the knee center.

(ii) Both knee flexors and extensors are active, and the flexors have 50 N.m torque acting at a 6 cm distance from the knee center. Consider the flexor tendon force to be acting vertically.

(i) From Table B.2 for frame 10 the knee moment is 103.1 N.m. The reaction force at the knee is −846 N in the vertical direction, and +141 N horizontally. This negative vertical reaction force means that the thigh is acting downward on the shank, producing compression equivalent to 846 ÷ (80 × 9.8) = 1.08 times body weight. The tension in the patella tendon resulting from the extensor moment = 103.1 ÷ .06 = 1718 N. This vertical force will be balanced by an equal and opposite joint compressive force such that the net bone-on-bone force is 1718 + 846 = 2564 N. This is 3.27 times body weight. The horizontal shear force of 141 N is still present.

(ii) The knee extensor moment will increase by 50 N.m if there is a 50 N.m antagonist in the flexors. Thus the extensor moment will now be 153.1 N.m. The patella tendon tension = 153.1 ÷ .06 = 2552 N. If the flexors act 6 cm from the knee center the force in the flexor tendon will be 50 ÷ .06 = 833 N. Both these forces will act vertically and be balanced by an equal compressive force acting across the joint articulating surfaces. The total compressive force will be 2552 + 833 + 846 = 4231 N, or 5.4 times body weight. Such a high bone-on-bone force which repeats itself stride after stride can be quite damaging to the articulating surfaces of the knee joint. This is a particular cause of knee problems in athletes, and of considerable pain in arthritic knee patients.

4.5   STUDENT PROBLEMS

1. For Frame 58 calculate the following and verify your answer with the results in the table indicated. Use appropriate kinematic data for the segments concerned, and employ the anthropometric data as provided in Table A.2.

   (a) Using kinematic data from Tables A.4 and A.5, calculate:

      (i) The ankle reaction forces (as they act on the foot).
      (ii) The net ankle muscle moment (as it acts on the foot).

   Verify your answers with Table B.1.

   (b) Using the answers from (a), (or if incorrect, Table B.1 results) plus kinematic data from Tables A.4 and A.6 calculate:

      (i) The knee reaction forces (as they act on the shank).

      (ii) The net knee muscle moment (as it acts on the shank).

   Verify your answers with Table B.2.

(c) Repeat (b) for the thigh segment using the appropriate results from (b) combined with data from Tables A.4 and A.7.

Verify your answers with Table B.3.

2. For frame 30 calculate the reaction forces and net muscle moments at the ankle, knee and hip. Proceed as in question 1, starting with the foot segment utilizing the ground reaction force data from Table A.10 plus the appropriate kinematic data.

3. (a) Plot the net ankle muscle moment (Table B.1) over the stance period (frames 3 to 42). Based on the magnitude and polarity of this curve discuss the function of the ankle muscles during each of these phases of stance:

    (i) Heel-contact to foot flat (frames 3 to 12).
    (ii) Foot flat to heel-off (frames 12 to 20).
    (iii) Heel-off to toe-off (frames 20 to 42).

(b) Plot the net knee muscle moment (Table B.2) over stance and, based on the magnitude and polarity of this curve, discuss the role of the knee muscles during the following phases of gait:

    (i) Heel-contact to heel-off (frame 3 to 20).
    (ii) Heel-off to toe-off (frames 20 to 42).
    (iii) Toe-off to mid-swing (frames 42 to 50).
    (iv) Mid-swing to heel-contact (frames 50 to 70).

4. Assume that the Achilles tendon acts parallel to the long axis of the tibia at a distance of 6 cm from the ankle joint.

(a) Calculate the force (magnitude and direction) acting on the foot through the Achilles tendon during maximum muscle moment of push-off.

(b) Assuming only plantarflexor activity at the time, calculate the magnitude of the ankle bone-on-bone forces.

(c) Assuming that the tibialis anterior is co-contracting at this time with 20 N.m moment of force and that its force acts parallel to the tibia 5 cm in front of the ankle joint, calculate the ankle bone-on-bone forces.

REFERENCES

Bresler, B. and Frankel, J. P. The forces and moments in the leg during level walking. *Trans. ASME* **72:**27–36, 1950.

Elftman, H. Forces and energy changes in the leg during walking. *Amer. J. Physiol.* **125**:339–356, 1939.

Morrison, J. B. The mechanics of muscle function in locomotion. *J. Biomech.* **3**:431–451, 1970.

Paul, J. P. Forces transmitted by joints in the human body. *Proc. Inst. Mech. Eng.* **18**(3):8–15, 1966.

Pedotti, A. A study of motor coordination and neuromuscular activities in human locomotion. *Biol. Cybernetics* **26**:53–62, 1977.

CHAPTER FIVE
Mechanical Work, Energy, and Power

One of the major comparative measures of human performance is efficiency. The athlete and coach, especially in longer distance events, are striving for that extra 1% that separates victory from defeat. The amputee, on the other hand, is not interested in fine improvements, but in major changes that will mean at the end of the day he is not excessively tired. All efficiency calculations involve some measure of mechanical output divided by a measure of metabolic input. This chapter concentrates on the calculation of the numerator of this equation, and also aims to clarify certain anomalies that are associated with the analysis of the mechanical output of the human muscular system. These anomalies are due to improper measurements or calculations, and even to erroneous definitions.

5.0 DEFINITIONS AND THEORY

5.0.1 Metabolic, Mechanical and Overall Efficiency

The purpose of muscles is to produce tension; as such, the measure of a muscle's ability to convert from metabolic energy to tension is its metabolic efficiency. A high metabolic efficiency does not necessarily mean that an efficient movement is taking place. A cerebral palsy patient whose muscles are very spastic is usually quite efficient in converting metabolic energy to muscle tension. However, such a patient usually has severe co-contractions and jerky movement, indicating that the pattern of muscle tension is quite uncoordinated. The ability of the central nervous system to control the tension patterns is what influences the *mechanical efficiency*. A skilled athlete will normally have a high mechanical efficiency as well as a high metabolic efficiency. Because it is virtually impossible to measure these efficiencies separately, we are left with the calculation of an *overall muscle efficiency* = net mechanical work ÷ net metabolic energy.

5.0.2   Net Mechanical Work—Internal versus External Work

Depending on the activity, we can be faced with the problem of measuring two types of work done by the muscles. The first work done by the muscles is to move the limb segments through some desired pattern. This is called internal work, and many movements such as walking and running involve only internal work. The second form of work done by the muscles is external to the body, such as lifting weights, pushing a car, or bicycling against an external ergometer load. The net mechanical work = internal work + external work. It should be noted that external work includes lifting one's own body weight to a new height. Thus running up a hill involves both external work and internal work.

5.0.3   Positive Work of Muscles

Positive work is that done during a concentric contraction, when the muscle moment acts in the same direction as the angular velocity of the joint. If a flexor muscle is causing a shortening we can consider the flexor moment to be positive and the angular velocity to be positive. The product of muscle moment and angular velocity is positive; thus power is positive, and is depicted in Figure 5.1a. Conversely, if an extensor muscle moment is negative and an extensor angular velocity is negative the product is still positive, as shown in Figure 5.1b. The integral of this power over the time of the contraction is the net work done by the muscle and represents generated energy transferred from the muscles to the limbs.

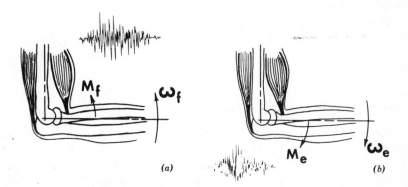

Figure 5.1   Positive work as defined by the net muscle moment and angular velocity.
a When a flexion moment acts while the forearm is flexing.
b When an extension moment acts during an extensor angular velocity. (By permission of *Physiotherapy Canada*.)

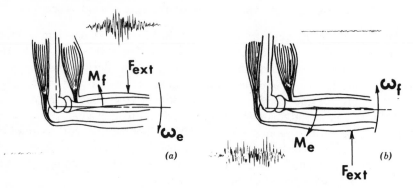

Figure 5.2 Negative work, as defined by net muscle moment and angular velocity.
a When an external force causes extension when the flexors are active.
b When an external force causes flexion in the presence of an extensor muscle moment.
(By permission of *Physiotherapy Canada.*)

5.0.4   Negative Work of Muscles

Negative work is that done during an eccentric contraction, when the muscle moment acts in the opposite direction to the movement of the joint. This usually happens when an external force, Fext, acts on the segment and is such that it creates a joint moment greater than the muscle moment. The external force could include gravitational or ground reaction forces. Using the polarity convention as described above we can see in Figure 5.2*a* we have a flexor moment (+ve) with an extensor angular velocity (−ve). The product yields a negative power, so that the work done during this angular change is negative. Similarly, when there is an extensor moment (−ve) during a flexor angular change (+ve), the product is negative (Figure 5.2*b*). Here the net work is being done by the external force on the muscles and represents a flow of energy from the limbs into the muscles (absorption).

5.0.5   Muscle Mechanical Power

The rate of work done by most muscles is rarely constant with time. Because of rapid time-course changes it has been necessary to calculate muscle power as a function of time (Elftman, 1939; Quanbury et al., 1975; Cappozzo et al., 1976; Winter and Robertson, 1978a). At a given joint, muscle power is the product of the net muscle moment and angular velocity.

$$Pm = Mj \; \omega j \quad \text{W} \tag{5.1}$$

where: *Pm* is the muscle power in watts.

Mj is the net muscle moment (N.m).

ωj is the joint angular velocity (rad/sec).

As has been described in the previous sections, *Pm* can be either positive or negative. During even the simplest movements the power will reverse sign several times. Figure 5.3 depicts the muscle moment, angular velocities, and muscle power as a function of time during a simple extension and flexion of the forearm. As can be seen the time course of *Mj* and *ωj* are roughly out of phase by 90°. During the initial extension there is an extensor moment and an extensor angular velocity as the triceps do positive work on the forearm. During the latter extension phase the forearm is decelerated by the biceps (flexor moment). Here the biceps are doing negative work (absorbing mechanical energy). Once the forearm is stopped it starts accelerating in a flexor direction still under the moment created by the biceps; the biceps are now doing positive work. Finally, at the end of the movement, the triceps decelerate the forearm as the extensor muscles lengthen; here, *Pm* is negative.

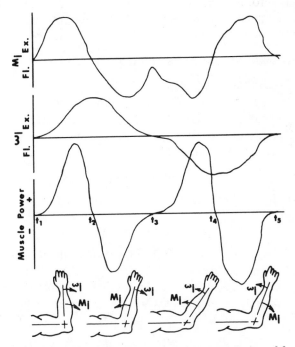

Figure 5.3 Sequence of events during simple extension and flexion of forearm. Muscle power shows two positive bursts alternating with two negative bursts.

5.0.6   Mechanical Work of Muscles

Until now we have used the terms power and work almost interchangeably. Power is the rate of doing work, thus, to calculate work done we must integrate power over a period of time. The product of power and time is work and it is measured in joules (1 J = 1 Wsec). If a muscle generates 100 W for 0.1sec the mechanical work done is 10 J. This means that 10 J of mechanical energy have been transferred from the muscle to the limb segments. As the example of Figure 5.3 showed, power is continuously changing with time. Thus the mechanical work done must be calculated from the time integral of the power curve. The work done by a muscle during a period t_1 to t_2 is:

$$Wm = \int_{t_1}^{t_2} Pm \; dt \quad J \tag{5.2}$$

In the example described the work done from t_1 to t_2 is positive, from t_2 to t_3 it is negative, from t_3 to t_4 is positive again, and during t_4-t_5 it is negative. If the forearm returned to the starting position, the net mechanical work done by the muscles is zero, meaning that the time integral of Pm from t_1 to t_5 is zero. It is therefore important to know the exact times when Pm is reversing polarities in order to calculate the total negative and total positive work done during the event.

5.0.7   Mechanical Work Done on an External Load

When any part of the body exerts a force on an adjacent segment or on an external body it can do work if there is movement. In this case work is defined as the product of the force acting on a body and the displacement of the body in the direction of the applied force. The work, dW, done when a force causes an infinitesimal displacement, dS is:

$$dW = F \; ds \tag{5.3}$$

Or the work done when F acts over a distance S_1, is:

$$W = \int_0^{S_1} Fds = FS_1 \tag{5.4}$$

If the force is not constant (which is more often the case) then we have two variables which change with time. Therefore, it is necessary to calculate the power as a function of time and integrate the power curve with respect to time to yield the work done. Power is the rate of doing work, or dW/dt.

$$P = \frac{dW}{dt} = F \; \frac{ds}{dt}$$

$$\text{or } P = \bar{F} \cdot \bar{V} \tag{5.5}$$

where: P is the instantaneous power in watts.
  $\bar{F}$ is the force in N.
  $\bar{V}$ is the velocity in m/sec.

Since both force and velocity are vectors, we must take the dot product, or the product of the force and the component of the velocity that is in the same direction as the force. This will yield:

$$P = FV \cos \theta \qquad (5.6)$$

where: $\cos \theta$ is the angle between the force and velocity vectors.

For the purpose of this initial discussion let us assume that the force and velocity are always in the same direction. Therefore $\cos \theta = 1$ and:

$$P = FV \quad \text{W}$$

$$W = \int_0^t P\,dt = \int_0^t FV\,dt \quad \text{J} \qquad (5.7)$$

Example 5.1.   A baseball is thrown with a constant accelerating force of 100 N for a period of 180 msec. The mass of the baseball is 1.0 kg, and it starts from rest. Calculate the work done on the baseball during the time of force application.

Solution $S_1 = ut + 1/2\ at^2$
  $u = 0$
  $a = F/m = 100/1.0 = 100$ m/sec^2
  $S_1 = 1/2 \times 100 \times (.18)^2 = 1.62$ m
  $W = \int_0^{S_1} F\,ds = FS_1 = 100 \times 1.62 = 162$ J

Example 5.2.   A baseball of mass = 1 kg is thrown with a force which varies with time, as indicated in Figure 5.4. The velocity of the baseball in the direction of the force is also plotted on the same timebase and was calculated from the time integral of the acceleration curve (which has the same numerical value as the force curve because the mass of the baseball is 1 kg). Calculate the instantaneous power to the baseball and the total work done on the baseball during the throwing period.

The peak power calculated here may be considered quite high, but it should be noted that this peak is of short duration. The average power for the throwing period is less than 500 W. In real-life situations it is highly unlikely that the force will ever be constant; thus instantaneous power will always need to be calculated.

When the baseball is caught, the force of the hand still acts against the baseball, but the velocity is reversed. The force and velocity vectors are now in the opposite direction. Thus the power is negative and the work done is also negative, indicating that the baseball is doing work on the body.

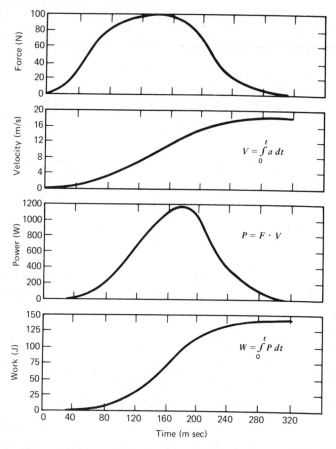

Figure 5.4 Forces, velocity, mechanical power, and work done on a baseball while being thrown. See text for details.

5.0.8 Mechanical Energy Transfer Between Segments

Each body segment exerts forces on its neighboring segments, and if there is a translational movement of the joints there is a mechanical energy transfer between segments. In other words, one segment can do work on an adjacent segment by a force-displacement through the joint center (Quanbury, et al., 1975). This work is in addition to the muscular work described in Sections 5.0.3–5.0.6. Equations 5.5 and 5.6 can be used to calculate the rate of energy transfer (i.e., power) across the joint center. Consider the situation in Figure 5.5 at the joint between two adjacent segments. Fj_1, the reaction force of segment 2 on segment 1 acts

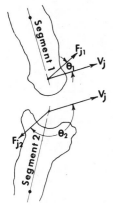

Figure 5.5 Reaction forces and velocities at a joint center during dynamic activity. The dot product of the force and velocity vectors is the mechanical power (rate of mechanical energy transfer) across the joint.

at an angle θ_1 from the velocity vector Vj. The product of $Fj_1Vj \cos \theta_1$ is positive, indicating that energy is being transferred into segment 1. Conversely, $Fj_2 Vj \cos \theta_2$ is negative, denoting a rate of energy outflow from segment 2. Since $Pj_1 = -Pj_2$, the outflow from segment 2 equals the inflow to segment 1.

This mechanism of energy transfer between adjacent segments is quite important in the conservation of energy of any movement because it is a passive process and does not require muscle activity. In walking this has been analyzed in detail (Winter and Robertson, 1978a). At the end of swing, for example, the swinging foot and leg lose much of its energy by transfer upward through the thigh to the trunk where it is conserved and converted to kinetic energy.

5.1 CAUSES OF INEFFICIENT MOVEMENT

It is often difficult as a therapist or coach to concentrate directly on efficiency. Rather, it is more reasonable to focus on the individual causes of inefficiency, and thereby automatically improve the efficiency of the movement. The four major causes of mechanical inefficiency (Winter, 1978b) will now be described.

5.1.1 Co-Contractions

Obviously, it is inefficient to have muscles co-contract because they fight against each other without producing a net movement. Suppose a certain movement can be accomplished with a flexor moment of 10 N.m.

The most efficient way to do this is with flexor activity only. However, the same movement can be achieved with 20-N.m flexion and 10-N.m extension, or with 40-N.m flexion and 30 N.m extension. In the latter case, there is an unnecessary 30 N.m moment in both the extensors and flexors. Another way to look at this situation is that the flexors are doing unnecessary positive work to overcome the negative work of the extensors.

Co-contractions occur in many pathologies, notably hemiplegia and spastic cerebral palsy. They also occur to a limited extent during normal movement when it is necessary to stabilize a joint, especially if heavy weights are being lifted or at the ankle joint during walking or running. At present the measurement of unnecessary co-contractions is only possible by monitoring the EMG activity of the antagonistic muscles. However, without an exact EMG calibration versus tension for each muscle, it is impossible to arrive at a quantitative measure of co-contraction.

5.1.2 Isometric Contractions Against Gravity

In normal dynamic movement there is minimal muscle activity that can be attributed to holding limb segments against the forces of gravity. This is because the momentum of the body and limb segments allows for a smooth interchange of energy. However, in many pathologies the movement is so slow that there are extended periods of time when limb segments or the trunk are being held in near-isometric contractions. Spastic cerebral palsy patients often stand with their knee flexed, requiring excessive quadriceps activity to keep them from falling down. Or, as seen in Figure 5.6, the crutch-walking cerebral palsy child holds her leg off the ground for a period of time prior to swing through.

5.1.3 Jerky Movements

Efficient energy exchanges are characterized by smooth-looking movements. A ballet dancer and a high jumper execute smooth movements for different reasons, one for artistic purposes, the other for efficient performance. Energy added to the body by positive work at one point in time is conserved, and very little of this energy is lost by muscles doing negative work. The jerky gait of a cerebral palsy child is quite the opposite. Energy added at one time is removed a fraction of a second later. The movement has a steady succession of stops and starts, and each of these bursts of positive and negative work has a metabolic cost. The energy cost due to jerky movements can be assessed by a total body energy analysis, which will be described later.

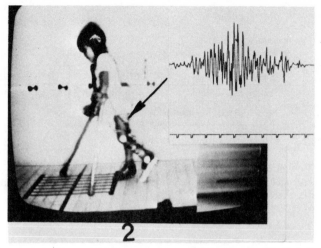

Figure 5.6 Example of "work" against gravity, one of the causes of inefficient movement. Here a cerebral palsy child holds her leg against gravity for an extended period prior to swinging through. (By permission of *Physiotherapy Canada*.)

5.1.4 Generation of Energy at One Joint and Absorption at Another

The least known and understood cause of inefficiency is when one muscle group at one joint does positive work at the same time as negative work is being done at others. Such an occurrence is really an extension of what occurs during a co-contraction (e.g., positive work being cancelled out by negative work). It is quite difficult to visualize when this happens. During normal gait it occurs during double support when the energy increase of the push-off leg takes place at the same time as the weight-accepting leg absorbs energy. Figure 5.7 shows this point in gait: the left leg push-off is (positive work) due primarily to plantarflexors; the right leg energy absorption (negative work) takes place in the quadriceps and tibialis anterior. There is no doubt that the instability of pathological gait is a major cause of this type of inefficient muscle activity.

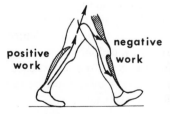

positive work

negative work

Figure 5.7 Example of a point in time during gait that positive work by the push-off muscles can be cancelled by negative work of the weight-accepting muscles of the contralateral leg. (By permission of *Physiotherapy Canada*.)

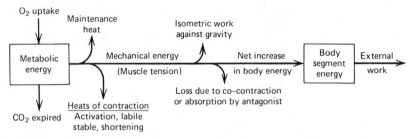

Figure 5.8   Flow of energy from metabolic level to external mechanical work. Energy is lost as heat associated with the contractile process or by inefficiencies after energy has been converted to mechanical energy.

5.1.5   Summary of Energy Flows

It is valuable to summarize the flows of energy from the metabolic level through to an external load. Figure 5.8 depicts this process schematically. Metabolic energy cannot be measured directly, but can be calculated indirectly from the amount of O_2 required or by the CO_2 given off. The details of these calculations and their interpretation are the subject of other textbooks and are beyond the scope of this book.

At the basal level (resting, lying down) the muscles are relaxed but still require metabolic energy to keep them alive. The measure of this energy level is called maintenance heat. Then as a muscle contracts it requires energy, which shows up as additional heat. Activation heat is that associated with the rate of buildup of tension within the muscle and is accompanied by an internal shortening of the muscle contractile elements. Stable heat is that heat which measures the energy required to maintain tension within the muscle. Labile heat is a third heat seen in isometric contractions, and is that not accounted for by either tension or rate of tension generation. The final heat loss is that associated with the actual shortening of the muscle under load.

Finally, at the tendon we see the energy in mechanical form. The muscle tension can contribute to four types of mechanical loads. It can be involved with a co-contraction, isometric "work" against gravity, or a simultaneous absorption by another muscle, or it can cause a net change in the body energy. In the latter case, if positive work is done the net body energy will increase; if negative work is done there will be a decrease. Finally, if the body is exerting forces on an external body, some of the energy may be transferred as the body performs external work.

5.2 FORMS OF ENERGY STORAGE

(i) *Potential Energy, P.E.*

Potential energy is the energy due to gravity, and therefore increases with height of the body above ground or above some other suitable reference datum.

$$P.E. = mgh \quad J \tag{5.8}$$

where: m is mass in kg

g is gravitational acceleration = 9.8 m/sec^2

h is height of center mass in m

With $h = 0$, the potential energy decreases to zero. However, the ground reference datum should be carefully chosen to fit the problem in question. Normally it is taken as the lowest point the body takes during the given movement. For a diver it could be the water level; for a person walking it would be the lowest point in the pathway.

(ii) *Kinetic Energy, K.E.*

There are two forms of kinetic energy: that due to the translational velocity and that due to rotational velocity.

$$Translational \ K.E. = 1/2 \ mv^2 \quad J \tag{5.9}$$

where: v is velocity of center of mass in m/sec

$$Rotational \ K.E. = 1/2 \ I\omega^2 \quad J \tag{5.10}$$

where: I is rotational moment of inertia in kg. m^2

ω is rotational velocity of segment in rad/sec

Note that these two energies increase as the (velocity)^2. The polarity or direction of the velocity is unimportant because (velocity)^2 is always positive. The lowest level of kinetic energy is therefore 0, when a body is at rest.

(iii) *Total Energy and Exchange Within a Segment*

As mentioned previously, energy of a body exists in three forms, so that the total energy of a body,

$$Es = P.E. + translational \ K.E. + rotational \ K.E.$$
$$= mgh + 1/2 \ mv^2 + 1/2 \ I\omega^2 \quad J \tag{5.11}$$

It is possible for a body to exchange energy within itself and still maintain a constant total energy.

Example 5.3. Suppose the baseball in Example 5.1 is thrown vertically. Calculate the potential and kinetic energies at the time of release, at maximum height, and when it reaches the ground. Assume it is released at a height of 2 m above the ground. Assume that the vertical accelerating force of 100 N is in excess of gravitational force.

At release, P.E. $= mgh = 1 \times 9.8 \times 2 = 19.6$ J

$$a = 100 \text{m/sec}^2 \text{ (as calculated previously)}$$

$$v = \int_0^{t_1} adt = at_1 = 100 \, t_1$$

$$t_1 = 180 \text{ msec}$$

$$v = 18 \text{ m/sec}$$

Translational K.E. $= 1/2 \, mv^2 = 1/2 \times 1.0 \times 18^2 = 162$ J

Note that this 162 J is equal to the work done on the baseball prior to release.

$$\text{Total energy} = \text{P.E.}(t_1) + \text{translational K.E. } (t_1)$$
$$= 19.6 + 162 = 181.6 \text{ J}$$

If we ignore air resistance, the total energy remains constant during the flight of the baseball, such that at t_2 when the maximum height is reached all the energy is P.E., and K.E. $= 0$.

$$\text{P.E. } (t_2) = 181.6 \text{ J}$$

This means that the baseball reaches a height such that

$$mgh_2 = 181.6 \text{ J}$$

$$h_2 = \frac{181.6}{1.0 \times 9.8} = 18.5 \text{ m}$$

At t_3 the baseball strikes the ground, and $h = 0$

$$\text{Thus} \quad \text{P.E. } (t_3) = 0$$

$$\text{K.E. } (t_3) = 181.6 \text{ J}$$

This mean that the velocity of the baseball is such that

$$1/2 \, mv^2 = 181.6 \text{ J}$$

$$\text{or} \qquad v = 19.1 \text{ m/sec}$$

This is slightly higher than the release velocity of 18 m/sec because the ball was released from 2 m above ground level.

5.2.1 Energy of a Body Segment, and Exchanges of Energy Within the Segment

Most body segments contain all three energies in various combinations at any point in time during a given movement. A diver at the top of his dive has considerable potential energy, and during the dive converts it to kinetic energy. Similarly, a boomerang when released has rotational and translational kinetic energy, and at peak height some of the translational kinetic has been converted to potential energy. At the end of its travel the boomerang will have regained most of its translational kinetic energy.

In a multisegment system such as the human body, the exchange of energy can be considerably more complex. There can be exchanges within a segment or between adjacent segments. A good example of energy exchange within a segment is during normal gait. The upper part of the body (head, arms, and trunk, H.A.T.) has two peaks of potential energy each stride—during mid-stance of each leg. At this time H.A.T. has slowed its forward velocity to a minimum. Then, as the body falls forward to the double support position, H.A.T. picks up velocity at the expense of a loss in height.

Example 5.4. Examine the kinematic data for H.A.T. (Table A.8) to see if there is any evidence of energy exchange within that segment during one stride.

Evidence of energy exchange should be seen from a plot of the horizontal velocity and vertical displacement of the C of G of H.A.T. (Figure 5.9). The potential energy, which varies as height, changes roughly as a sinusoidal wave, with a minimum during double support and reaching a maximum during mid-stance. The forward velocity is almost 180° out of phase, with peaks approximately during double support and minimums during mid-stance.

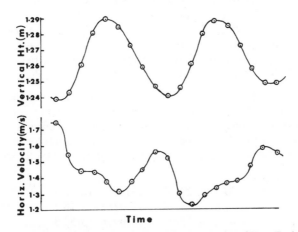

Figure 5.9 Plot of vertical displacement and horizontal velocity of H.A.T. shows evidence of energy exchange within the upper part of the body during gait.

Example 5.5. From the data in Table A.6 calculate the energy components of the shank during frame 55.

$$m = 3.72 \text{ kg} \qquad\qquad I = .064 \text{ kg} \cdot \text{m}^2$$
$$g = 9.8 \text{ m/sec}^2 \qquad\qquad \omega = 5.252 \text{ sec}^{-1}$$
$$h = y = .400 \text{ m}$$
$$Vx = -0.26 \text{ m/sec} \qquad V = \sqrt{Vx^2 + Vy^2} = 2.655 \text{ m/sec}$$
$$Vy = 2.642 \text{ m/sec}$$
$$Es = mgh + 1/2\ mv^2 + 1/2\ I\omega^2$$
$$= 3.72 \times 9.8 \times .4 + 1/2 \times 3.72 \times (2.655)^2 +$$
$$1/2 \times .064 \times (5.252)^2$$
$$= (14.58 + 13.11 + .88) \text{ J}$$
$$= 28.57 \text{ J}$$

Example 5.6. By visually scanning the energies of the shank (Table B.5) discuss:

(i) The relative importance of each of the three energy components during one stride.

(ii) Any evidence of energy exchange (an exchange occurs when one energy component increases while one of the other components decreases).

(i) The potential energy drops to 13.02 J during mid-stance (frame 20) and increases to 15.18 joules during mid-swing (frame 49). Similarly, the translational kinetic energy has a minimum of .09 J during stance (frame 17) and a maximum of 13.63 J during swing (frame 51). The rotational kinetic energy changes only slightly during the stride (from 0 to .95 J). Therefore the translational kinetic energy is most important, followed by the potential component. Rotational energy is relatively unimportant here, but during running it becomes more important than the potential energy changes.

(ii) There is little evidence of energy exchange within the shank, because all energy components increase during swing and decrease during stance. However, if all components were 100% in phase there would be zero exchange; the change in total segment energy would be equal to the sum of the component changes. The change in total energy,

$$\Delta Es = 29.30 - 13.14 = 16.16 \text{ J}$$
$$\Delta Ep = 15.18 - 13.02 = 2.16 \text{ J}$$
$$\Delta Ekt = 13.63 - .09 = 13.54 \text{ J}$$
$$\Delta Ekr = .95 - 0 = .95 \text{ J}$$

Since $\Delta Ep + \Delta Ekt + \Delta Ekr = 16.65$ J it can be said that $16.65 - 16.16 = 0.49$ J exchanged during the stride. Thus the shank is a highly nonconservative system.

5.2.2   Exact Formula for Energy Exchange Within Segments

The example just discussed, where the changes in energy over a given cycle were calculated, illustrates simple energy exchanges. If the individual energy components have several maximums and minimums we must calculate the sum of the absolute energy changes over the time period. The work done by the segment W_s, during the time of N sample periods (e.g., N movie frames) is:

$$W_s = \sum_{i=1}^{N} \left| \Delta E_s \right| \text{ J} \qquad (5.12)$$

If we assume no energy exchanges between any of the three components (Norman et al., 1976), the work done by the segment during the N sample periods is:

$$W_s' = \sum_{i=1}^{N} \left(\left| \Delta E_p \right| + \left| \Delta E_{kt} \right| + \left| \Delta E_{kr} \right| \right) \text{ J} \qquad (5.13)$$

Therefore, the energy conserved within the segment, W_c, during the time is:

$$W_c = W_s' - W_s \text{ J} \qquad (5.14)$$

The percentage energy conservation C_s during the time of this event is:

$$C_s = \frac{W_c}{W_s'} \times 100\% \qquad (5.15)$$

If $W_s' = W_c$, all three energy components are in phase (they have exactly the same shape and have their minimums and maximums at the same time), and then there is no energy conservation. Conversely, as demonstrated by an ideal pendulum, if $W_s = 0$ then 100% of the energy is being conserved.

5.2.3   Total Energy of a Multi-Segment System

As we proceed with the calculation of the total energy of the body we merely sum the energies of each of the body segments at each point in time (Bresler et al, 1951; Ralston and Lukin, 1969; Winter et al., 1976). Thus the total body energy, E_b, at a given time is:

$$E_b = \sum_{i=1}^{B} E_{si} \text{ J} \qquad (5.16)$$

where: E_{si} is the total energy of the ith segment at that point in time.
$\quad\quad\ B$ is the number of body segments.

The individual segment energies continuously change with time, so it is not surprising that the sum of these energies will also change with time. Before we can appreciate the significance of these changes we should look at a simple pendulum system.

5.2.4 Positive and Negative Work of the Total Body

5.2.4.1 Illustrative Example of a Pendulum. Consider an ideal (lossless) pendulum in which 100% energy exchange is known to occur. At the top of its swing the pendulum has zero kinetic energy and maximum potential energy. At the bottom of the swing the reverse is true. Figure 5.10 shows the two conditions.

$$\text{Potential energy} + \text{kinetic energy} = \text{constant}$$
$$mgh + 1/2\ mv^2 = k$$

If h_o is the maximum height, then

$$mgh_o + 0 = k$$
$$mg(h_o - h) = 1/2\ mv^2$$
$$\text{or } v = \sqrt{2g\ (h_o - h)}$$

A plot of the individual and total energies versus time is shown in Figure 5.11. Since this is a 100% conservative system of total energy, it is always constant. A change in the total energy of any system can only occur if energy is transferred to or from an adjacent body, or, as will now be seen, gained or lost by muscle activity.

5.2.4.2 Pendulum System With Muscles. Consider, now, a simple muscular system which can be represented by a pendulum mass with a pair of antagonistic muscle groups, m_1 and m_2, crossing a simple hinge joint. Figure 5.12 shows such an arrangement along with a time history of the total energy of the system. At t_1 the segment is rotating coun-

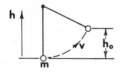

Figure 5.10 Pendulum swinging with a velocity, v, and height, h.

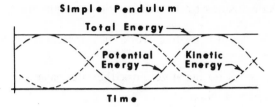

Figure 5.11   Exchange of kinetic and potential energy in a swinging frictionless pendulum. Total energy of the system is constant, indicating that no energy is being added or lost. (By permission of *Physiotherapy Canada.*)

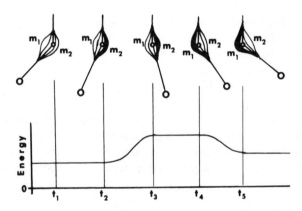

Figure 5.12   Pendulum system with muscles. When positive work is done, the total energy increases; when negative work is done, the total energy decreases. (By permission of *Physiotherapy Canada.*)

terclockwise at ω_1 rad/sec. No muscle activity occurs until t_2, when m_2 contracts. Between t_1 and t_2 the normal pendulum energy exchange takes place and the total energy remains constant. However, between t_2 and t_3 muscle m_2 causes an increase in both kinetic and potential energies of the segment. The muscle moment has been in the same direction as the direction of rotation, so positive work has been done by the muscle on the limb segment, and the total energy of the limb has increased. Between t_3 and t_4 both muscles are inactive, and the total energy remains at the higher but constant level. At t_4 muscle m_1 contracts to slow down the segment. Energy is lost by the segment and is absorbed by m_1. This is negative work being done by the muscle, because it is lengthening during its contraction. Thus at t_5, the segment has a lower total energy than at t_4.

The major conclusions to draw from this example are:

(i) When muscles do positive work they increase the total body energy.

(ii) When muscles do negative work they decrease the total body energy.

(iii) During cyclical activity such as level running, the net energy change per stride equals zero; thus the positive work done per stride equals negative work done per stride.

5.2.5   Total Body Work; Energy Exchanges Between Segments

The total of the positive and negative work done by a body system can be calculated from the total body energy curve, E_b (see Equation 5.16). The work done by the body, W_b, during N sample time periods is:

$$W_b = \sum_{i=1}^{N} |\Delta E_b| \quad \text{J} \tag{5.17}$$

Note the similarity of this formula with Equation 5.12, where we were calculating the total work done by a given segment. We can also calculate the amount of work that would have been done by the body assuming there was no energy exchange between segments. Designate this W_b^1.

$$W_b^1 = \sum_{i=1}^{B} \sum_{j=1}^{N} |\Delta E_{si}| \tag{5.18}$$

where: ΔE_{si} is the change in energy of the ith segment during the jth sample time
B is the number of body segments.

We can now calculate the energy that was actually conserved between all body segments:

$$W_{cb} = W_b^1 - W_b \quad \text{J} \tag{5.19}$$

The percentage of conservation between body segments, C_b, during the time of the event is:

$$C_b = \frac{W_{cb}}{W_b^1} \times 100\% \tag{5.20}$$

If $W_b^1 = W_b$, then all the segments are increasing and decreasing in energy at the same time and no energy exchange is taking place between segments.

Conversely, if $W_b = 0$, there is complete energy exchange at all times during the movement. This latter condition can actually occur during free flight, and could also be demonstrated in a frictionless puppet.

5.3 OVERALL EFFICIENCY OF HUMAN MOVEMENT

In the assessment of the energetics and efficiency of human movement it is important to have an accurate measure of the mechanical work done. At present there is considerable confusion and error in the techniques and definitions that are being used.

The first major error is in the assumption that the trajectory of the body's center of gravity contains the necessary information to calculate the mechanical energy of the body (Cavagna and Margaria, 1966). This error is particularly important in symmetrical or reciprocal movements such as walking or running. Symmetrical movements of limb segments in opposite directions does not necessarily result in a C of G change, yet there have been potential and kinetic energy changes in each of the moving limb segments.

The second problem is the anomalous situation associated with the efficiency calculation. Work physiologists classically define efficiency as work done by the body divided by metabolic cost. As such, exercise situations are arranged to ensure that external work is done: bicycling an ergometer, walking up a treadmill or steps, or lifting weights (Abbot et al., 1952; Cotes and Meade, 1960; Whipp and Wasserman, 1969; Pugh, 1971; Gaesser and Brooks, 1975). But what about the work done in moving the limbs themselves? We have defined such work as internal work, and this involves both positive and negative work. The present definition gives a positive efficiency when walking up a grade because of the mechanical work done to increase the potential energy of the body itself. However, this definition is anomalous when considering level walking or running. Here, at the end of each stride the body returns to the same energy level, and the calculated external work is zero. Does this mean that the efficiency is zero? Also, if we walk downhill the external work is actually negative; here, the classical definition implies that the efficiency is negative, or less than zero! Obviously, there is something wrong with the present definition. A simple modification can overcome this erroneous situation if the definition of efficiency, η, is modified as follows:

$$\eta = \frac{\text{external} + \text{internal mechanical work}}{\text{metabolic cost}} \times 100\% \quad (5.21)$$

The major problem is to calculate the internal mechanical work.

5.3.1   Internal Mechanical Work Done During a Movement

The time-course curve of total energy of a body system during the period of a movement is the key to the energy requirements. This curve is the sum of all energy components of all body segments and therefore reflects energy conservation within and between segments as well as positive or negative work (Winter, 1979). As has been discussed in Section 5.2.5 we already have a measure of the total mechanical work done by the body which takes into account all these energy exchanges. W_b (equation 5.17) gives us this value. It is the net internal work done by the limb segments themselves.

Example 5.7.   The total energy of the body is given in Table B.8. The integrated absolute energy changes are shown in the last column, commencing with frame 1. Determine the internal work done during the first step, and during the second step.

The work done during the stride (frame 1 to frame 69) is 166.54 J. During the first step (frames 1 to 36) it is 110.01 J. During the second step (frames 36 to 69) the work done is 166.54 − 110.01 = 56.53 J. The difference in the work done per step appears to be due to the fact that the person is slowing down, as is indicated in the decrease in forward velocity (see Figure 5.9).

If we now look at Equation 5.21 we note that the internal work term includes both positive and negative work, and the external work term is always positive work. The metabolic cost is a single measure and therefore contains all the metabolic requirements above the basal level, including those required for negative work. The student should be alerted to the anomalous results reported in the literature. Most researchers ignore the negative work component, which can cause serious problems in certain movements because the negative work can be equal to the positive work. A hypothetical example will illustrate the point. Suppose in level walking the positive and negative work per stride was 60 J each, and the metabolic cost was 300 J. Using positive work only, the efficiency would be 20%; using both, the efficiency would double to 40%. It is known that negative work is more efficient than positive work (Abbot et al., 1952). If we assume that negative work is twice as efficient as positive, we would now calculate positive work efficiency to be 30% and negative work efficiency to be 60%. This means that 200 J of the metabolic cost went into positive work and 100 J into the eccentric contractions.

5.4 APPLICATION OF ENERGY ANALYSES TO TRAINING, THERAPY, AND DIAGNOSTICS

The detailed diagnostic information present in the various energy curves is valuable to coaches, therapists, and clinicians to improve the performance of the athlete or patient. The overall work done during a given period of time is an excellent measure with which to monitor improvement or lack of improvement. The detailed segment energies can be used to diagnose the cause of an inefficiency.

5.4.1 Monitoring of Total Mechanical Work

The total mechanical energy curve, Es, as calculated by equation 5.11, summarizes the net positive and negative work done to accomplish any movement task. Inefficiencies and awkwardness show up as an above-normal change in Es and an increase in the total internal work, W_s (Equation 5.12). Such an assessment has been done on walking normals (Winter, 1979) and it was found that the average value of W_s was 1.1 J/kg.m. This means that a 70-kg person would do about 77,000 J of mechanical work to walk 1000 m. The same measure of mechanical work per unit of body mass per distance walked has also been used to monitor the progress of patient treatment (Winter, 1978a). There is no reason to believe that such a measure would not also serve to assess the efficiency of runners in a variety of events.

5.4.2 Detailed Diagnostics

If we find from the calculation of W_s that an inefficient movement exists then, we can examine the Es curve in detail to see when and what segment was responsible. Such an assessment has been reported (Winter, 1978b) in a study of the gait of a hip disarticulation amputee. This study revealed some interesting and unexpected results. The energy required to swing the prosthesis through was negligible, mainly because of its low mass and slow swing velocity. This finding is quite different from normal gait, where 70% of the total body energy is required to accelerate and decelerate the lower limbs during swing (Ralston and Lukin, 1969; Winter et al., 1976). It was found with this high level amputee that the large majority of his mechanical energy was required to lift his trunk so that he could swing through his prosthetic leg without danger of stubbing his foot on the ground. The solution to this amputee's problem was not revised gait training; rather, a major redesign of his prosthesis was indicated.

5.5 STUDENT PROBLEMS

1. **(a)** Calculate the potential and kinetic (translational and rotational) energies for the right foot for frame 55. Compare their relative importance and compare the horizontal translational kinetic energy with the vertical component at that time.

   (b) Calculate the total energy of the foot for frame 55 and compare with that shown in Table B.4.

   (c) Plot the three component energies and total energy for one stride and estimate how much energy is conserved within this segment during the stride. Express this as a percentage of the total possible energy change during stride.

   (d) Repeat part **(c)** for the thigh energies as calculated in Table B.6.

2. **(a)** Calculate the potential and kinetic (translational and rotational) energies for H.A.T. for frame 55.

   (b) Plot the three component energies and total energy for one stride and estimate how much energy is conserved within H.A.T. during the stride. Express this as a percentage of the total possible energy change during the stride.

   (c) From the curves plotted in part **(b)** compare the relative importance of each of the components.

3. Predict for a runner whose average velocity is three times that of the subject walking, what would happen to each of the energy components of each body segment. Assume that the stride length increases 50% in running, and that double support decreases from 20% of stride period to -20% (free flight now occupies 20% of the time).

4. **(a)** Given that the number of frames per step = 35, shift the results of the energy analyses for the right leg to create a set of data for a left leg (assuming symmetrical gait). For example, the energies for the left leg (frame 1) = energies calculated for the right leg (frame 36). Calculate the energy of the body for frame 21 by first determining the energies of the right leg, H.A.T., and left leg. Compare with that calculated in Table B.8.

   (b) Commencing at frame 1, plot the energy of the right and left legs on the same graph with the total energy of H.A.T. Discuss the relative importance of the energy of the legs compared with that of H.A.T.

REFERENCES

Abbot, B. C., Bigland, B. and Ritchie, J. M. The physiological cost of negative work. *J. Physiol.* **117**:380–390, 1952.

Bresler, B. and Berry, F. Energy levels during normal level walking. *Report of Prosthetic Devices Research Project*, University of California, Berkeley, May 1951.

Cappozzo, A., Figura, F. and Marchetti, M. The interplay of muscular and external forces in human ambulation. *J. Biomech.* **9**:35–43, 1976.

Cavagna, G. A. and Margaria, R. Mechanics of walking. *J. Appl. Physiol.* **21**:271–278, 1966.

Cotes, J. and Meade, F. The energy expenditure and mechanical energy demand in walking. *Ergonomics.* **3**:97–119, 1960.

Elftman, H. Forces and energy changes in the leg during walking. *Amer. J. Physiol.* **125**:339–356, 1939.

Gaesser, G. A. and Brooks, G. A. Muscular efficiency during steady rate exercise: effects of speed and work rate. *J. Appl. Physiol.* **38**:1132–1139, 1975.

Hill, A. V. Production and absorption of work by muscle. *Science.* **131**:897–903, 1960.

Morrison, J. B. Mechanics of muscle function in locomotion. *J. Biomech.* **3**:431–451, 1970.

Norman, R. W., Sharratt, M., Pezzack, J. and Noble, E. A re-examination of mechanical efficiency of horizontal treadmill running. *Biomechanics V-B*, Ed. by P. V. Komi, Baltimore, University Park Press, pp 87–93, 1976.

Pugh, L. Influence of wind resistance in running and walking and the mechanical efficiency of work against horizontal or vertical forces. *J. Physiol.* **213**:255–276, 1971.

Quanbury, A. O., Winter, D. A., and Reimer, G. D. Instantaneous power and power flow in body segments during walking. *J. Human Movement Studies.* **1**:59–67, 1975.

Ralston, H. J. and Lukin, L. Energy levels of human body segments during level walking. *Ergonomics.* **12**:39–46, 1969.

Whipp, B. J. and Wasserman, K. Efficiency of muscular work. *J. Appl. Physiol.* **26**:644–648, 1969.

Winter, D. A., Quanbury, A. O. and Reimer, G. D. Analysis of instantaneous energy of normal gait. *J. Biomech.* **9**:253–257, 1976.

Winter, D. A. and Robertson, D. G. Joint torque and energy patterns in normal gait. *Biol. Cybernetics* **29**:137–142, 1978a.

Winter, D. A. Energy assessments in pathological gait. *Physiotherapy Canada* **30**:183–191, 1978b.

Winter, D. A. A new definition of mechanical work done in human movement. *J. Appl. Physiol.* **46**:79–83, 1979.

CHAPTER SIX
Muscle Mechanics

The most intriguing and challenging area of study in biomechanics is probably the muscle itself. It is the "living" part of the system; the neural control, metabolism, and biomechanical characteristics of muscle are the subject of continuing research. The purpose of this chapter is to report the state of knowledge regarding the biophysical characteristics of individual motor units, connective tissue, and the total muscle itself. Some time will be spent on describing the characteristics of the individual units and how these characteristics influence the biomechanical function of the overall muscle.

6.0 THE MOTOR UNIT

The smallest subunit that can be controlled is called a motor unit because it is separately innervated by a motor axon. Neurologically, the motor unit consists of a synaptic junction in the ventral root of the spinal cord, a motor axon, and a motor end plate in the muscle fibers. Under the control of the motor unit are as few as three muscle fibers or as many as 2000, depending on the fineness of the control required (Feinstein et al., 1955). Muscles of the fingers, face, and eyes have a small number of shorter fibers in a motor unit, while the large muscles of the leg have a large number of long fibers in their motor units. A muscle fiber is about 100 μm in diameter, consisting of fibrils about 1 μ in diameter. Fibrils in turn consist of filaments about 100 Å in diameter. Electron micrographs of fibrils show us the basic mechanical structure of the interacting actin and myosin filaments. In the schematic diagram shown in Figure 6.1 we see the darker and wider myosin protein bands interlaced with the lighter and smaller actin protein bands. The space between them consists of a cross-bridge structure, and it is here that the tension is created and the shortening or lengthening takes place. The term contractile element is used to describe this part of the muscle that generates the tension. The basic length of the myofibril is the distance between the Z lines and is called the sarcomere length. It can vary from 1.5 μm at full shortening through 2.5 μm at resting length, to about 4.0 μm at full lengthening.

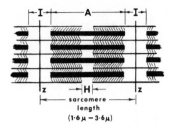

Figure 6.1 Basic structure of the muscle contractile element showing the Z lines and sarcomere length. Wider dark myosin filament interacts across cross-bridges (cross-hatched lines) with the narrower actin filament. Darker and lighter bands (*A*, H and *I*) are shown.

The structure of the muscle is such that many filaments are parallel and many sarcomere elements are in series to make up a single contractile element. Consider a motor unit of cross-sectional area = 0.1 cm², and a resting length of 10 cm. The number of sarcomere contractile elements in series would be 10 cm/2.5 μm = 40,000, and the number of filaments (each with an area = 10^{-8} cm²) in parallel would be 0.1/10^{-8} = 10^7. Thus the number of contractile elements of sarcomere length packed into this motor unit would be 4 × 10^{11}.

The active contractile elements are contained within another fibrous structure of connective tissue called fascia. This tissue encloses the muscles, separating them into layers and groups and ultimately connecting them to the tendon at either end. The mechanical characteristics of connective tissue are important in the overall biomechanics of the muscle. Some of the connective tissue is in series with the contractile element, while some is in parallel. The effect of this connective tissue has been modelled as springs and viscous dampers, and is discussed in detail in Section 6.4.

6.1   RECRUITMENT OF MOTOR UNITS

Each muscle has a finite number of motor units, each of which is controlled by a separate nerve ending. Excitation of each unit is an all-or-nothing event. The electrical indication is a motor unit action potential; the mechanical result is a twitch of tension. An increase in tension can therefore be accomplished in two ways: by an increase in stimulation rate for that motor unit or by the excitation (recruitment) of an additional motor unit. In Figure 6.2 we see the EMG of a needle electrode in a muscle as the tension was gradually increased. The upper tracing shows one motor unit firing, the middle trace two motor units, and the lower trace three units. Initially, muscle tension increases because the "firing" rate increases. At a certain tension the second motor unit was recruited,

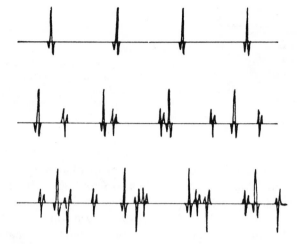

Figure 6.2 EMG from an indwelling electrode in a muscle as it begins to develop tension. The smallest motor unit is first to be recruited, its rate increases to its maximum, and then a second and third motor unit are recruited. Each m.a.p. has a characteristic shape at a given electrode.

and further tension increases are then accomplished by increases in rate of the second motor unit plus possible further increases in the rate of the first unit. As each unit has a maximum firing rate it is possible that this maximum rate is reached just as a new unit is recruited. When tension is reduced, the reverse process occurs. The firing rate of the recruited units decreases until the minimum rate for the last-recruited unit is reached, at which point the unit drops out. Each unit usually drops out in the reverse order to which it was recruited. A mathematical model describing this phenomenon has been developed by Wani and Guha (1975) and Milner-Brown and Stein (1975).

6.1.1   Size Principle

Considerable research and controversy has taken place over the past two decades over how the motor units are recruited. Which ones are recruited first? Are they always recruited in the same order? It is now generally accepted that they are recruited according to the *size principle,* (Henneman, 1974a) which states that the size of the newly recruited motor unit increases with the tension level at which it is recruited. This means the smallest unit is recruited first and the largest unit last. In this manner low tension movements can be achieved in finely graded steps. Conversely, those movements requiring high forces but not needing fine

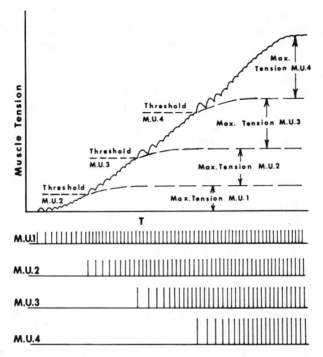

Figure 6.3 Illustration of size principle of recruitment of motor units. Smaller motor units are recruited first; successively larger units begin firing at increasing tension levels. In all cases the newly recruited unit fires at a base frequency, then increases to a maximum.

control are accomplished by recruiting the larger motor units. Figure 6.3 depicts a hypothetical tension curve resulting from successive recruitment of several motor units. The smallest motor unit (m.u.#1) is recruited first, usually at an initial frequency of about 4 Hz. Tension increases as m.u. 1 fires more rapidly, until a certain tension is reached at which m.u. 2 is recruited. Here, m.u.#2 starts firing at a low rate and further tension is achieved by the increased firing of both #1 and #2. At a certain tension m.u. 1 will reach its maximum firing rate (15 to 60 Hz) and will therefore be generating its maximum tension. This process of increasing tension, reaching new thresholds, and recruiting another larger motor unit continues until maximum voluntary contraction is reached (not shown in Figure 6.3). At that point all motor units will be firing at their maximum frequencies. Tension is reduced by the reverse process: successive reduction of firing rates and dropping out of the larger units first.

The muscle action potential (m.a.p.) increases with the size of the motor unit with which it is associated (Milner-Brown and Stein, 1975). The reason for this appears to be twofold. The larger the motor unit the larger the motoneuron that innervates it, and the greater the depolarization potentials associated with the motor end plate. The greater the mass of the motor unit, the greater the voltage changes seen at a nearby electrode. However, it is never possible from a given recording site to predict the size of a motor unit because the m.a.p. also decreases with the distance between the recording site and the electrode. Thus a large motor unit located a distance from the electrode may produce a smaller m.a.p. than that produced by a small motor unit directly beneath the electrode.

6.1.2   Types of Motor Units—Fast and Slow Twitch Classification

There have been many criteria and varying terminologies associated with the types of motor units present in any muscle (Henneman, 1974b). Biochemists have used metabolic or staining measures to categorize the fiber types. Biomechanics researchers have used force (twitch) measures (Milner-Brown et al., 1973b), and electrophysiologists have used EMG indicators (Warmolts and Engel, 1973; Milner-Brown et al., 1975). The smaller, slow twitch motor units have been called tonic units. His-tochemically they are the smaller units (Type I) and metabolically they have fibers rich in mitochondria, are highly capillarized, and therefore, have a high capacity for aerobic metabolism. Mechanically, they produce twitches with a low peak tension with a long time to peak (60 to 120 msec). The larger, fast twitch motor units are called phasic (Type II) units. They have little mitrochondria, are poorly capillarized, and therefore, rely on anaerobic metabolism. The twitches have larger peak tensions in a shorter time (10 to 50 msec). Figure 6.4 is a typical histochemical stain of fibers of a muscle which contains both slow and fast twitch fibers. An ATPase type stain was used here, so slow twitch fibers appear dark and the fast twitch fibers appear light. If an indwelling microelectrode were present in the area of these fibers the muscle action potential from these darker slow twitch fibers would be smaller than from the lighter-stained fast twitch fibers.

In spite of the two compartment classifications described above, there is growing evidence (Milner-Brown et al., 1973b) that the motor units controlled by any motoneuron pool form a continuous spectrum of sizes and excitabilities.

6.2   FORCE-LENGTH CHARACTERISTIC OF MUSCLES

As indicated in Section 6.0, the muscle consists of an active element, called the contractile element, and passive connective tissue. The net

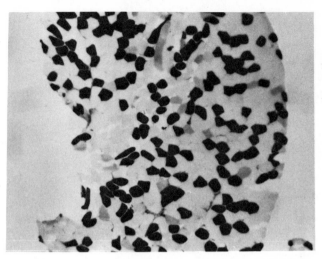

Figure 6.4 Histochemical stain showing dark slow-twitch fibers and light fast-twitch fibers. A myofibrillar ATPase stain, pH 4.3, was used to stain the vastus lateralis of a female volleyball player. (By kind permission of Professor J. A. Thomson, University of Waterloo, Waterloo, Canada.)

force-length characteristic of a muscle is a combination of the force-length characteristics of both active and passive elements.

6.2.1 Force-Length Curve of the Contractile Element

The key to the shape of the force-length curve is the changes of the structure of the myofibril at the sarcomere level (Gordon et al., 1966). In Figures 6.5 a schematic representation of the myosin and actin crossbridge arrangement is shown. At resting length, about 2.5. μm, there are a maximum number of cross-bridges between the filaments, and therefore, maximum tension is possible. As the muscle lengthens the filaments are pulled apart and the number of cross-bridges reduces and tension decreases. At full length, about 4.0 μm, there are no cross-bridges and the tension reduces to zero. As the muscle shortens to less than resting length there is an overlapping of the cross-bridges and an interference takes place. This results in a reduction of tension that continues until a full overlap occurs, at about 1.5 μm. The tension doesn't drop to zero, but is drastically reduced by these interfering elements.

6.2.2 Influence of Parallel Connective Tissue

The connective tissue that surrounds the contractile element influences the force-length curve. It is called the parallel elastic component, and it

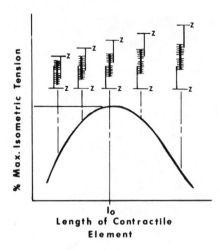

Figure 6.5 Tension produced by a muscle as it changes length about its resting length, l_0. Drop of tension on either side of maximum can be explained by interactions of cross-bridge attachments in the contractile element.

acts much like an elastic band. When the muscle is at resting length or less the parallel elastic component is in a slack state with no tension. As the muscle lengthens the parallel element is no longer loose, so tension begins to build up, slowly at first, and then more rapidly. Unlike most springs, which have a linear force-length relationship, the parallel element is quite nonlinear. In Figure 6.6 we see the force-length curve of this element, Fp, combined with that of the overall contractile component, Fc. If we sum the forces from both elements we see the overall muscle force-length characteristic, Ft. The force-length curve typically presented is usually for a maximal contraction. The passive force of the parallel element, Fp, is always present, but the amount of active tension in the contractile element at any given length is under voluntary control. Thus the overall

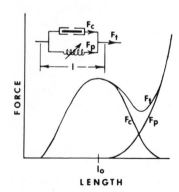

Figure 6.6 Contractile element producing maximum tension, F_c, along with the tension, F_p, from the parallel elastic element. Tendon tension is $F_t = F_c + F_p$.

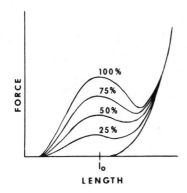

Figure 6.7   Tendon tension resulting from various levels of muscle contraction.

force-length characteristic is a function of percent of excitation, as seen in Figure 6.7.

The student can demonstrate the drop of tension at either end of the force-length curve by two simple experiments. The hamstrings, as a two joint muscle, can be made to shorten as follows: the person stands on one leg, leaning backward with the swing hip fully extended, then contracts the hamstrings to flex the leg. He will feel the tension decrease drastically when the muscle shortens before the knee is completely flexed. The converse situation can be realized if the person attempts to extend the hip joint while the knee is fully extended.

6.2.3   Series Elastic Tissue

All connective tissue in series with the contractile component, including the tendon, is called the series elastic element. Under isometric contractions it does not have any influence on the force-length characteristic. However, during dynamic situations the series elastic element, in conjunction with viscous components, does influence the time-course of the muscle tension.

During isometric contractions the series elastic component is under tension and therefore is stretched a finite amount. Because the overall length of the muscle is kept constant, the stretching of the series elastic element can only occur if there is an equal shortening of the contractile element itself. This is described as internal shortening. Figure 6.8 illustrates this point at several contraction levels. Although the external muscle length, L, is kept constant, the increased tension from the contractile element causes the series elastic element to lengthen by the same amount as the contractile element shortens internally. The amount of internal

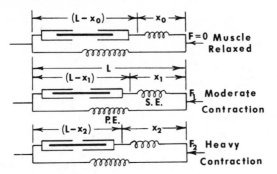

Figure 6.8   Introduction of the series elastic (S.E.) element. During isometric contractions the tendon tension reflects a lengthening of the series element and an internal shortening of the contractile element.

shortening from rest to maximum tension is only a few percent of the resting length.

Experiments to determine the force-length characteristics of the series element can only be done on isolated muscle and require dynamic changes of force or length. A typical experimental setup is shown in Figure 6.9. The muscle is stimulated to a certain level of tension while held at a certain isometric length. The load (tension) is suddenly dropped to zero by releasing one end of the muscle. With the force suddenly removed, the series elastic element, which is considered to have no mass, suddenly shortens to its relaxed length. This sudden shortening can be recorded, and the experiment repeated for another force level until a force-shortening curve can be plotted. During the rapid shortening period (about 2 msec) it is assumed that the contractile element has not had time to change length even though it is still generating tension. Two other

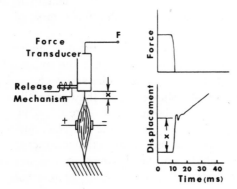

Figure 6.9   Experimental arrangement to determine the spring constant of the series elastic element. Muscle is stimulated; after the tension builds up in the muscle, the release mechanism activates, allowing an almost instantaneous shortening, x, while the force is recorded on a force transducer.

precautions must be observed when conducting this experiment. First, the isometric length prior to the release must be such that there is no tension in the parallel elastic element. This will be so if the muscle is at resting length or shorter. Second, the system that records the sudden shortening of the series elastic element must have negligible mass and viscosity. Because of the high acceleration and velocity associated with the rapid shortening the resisting force of the attached recording transducer should be negligible. If such conditions are not possible a correction should be made to account for the mass or viscosity of the transducer.

6.3   FORCE-VELOCITY CHARACTERISTICS

The previous section was concerned primarily with isometric contractions, and most physiological experiments are conducted under such conditions. However, movement cannot be accomplished without a change in muscle length. Alternate shortening and lengthening occurs regularly during any given movement, so it is important to see the effect of muscle velocity on muscle tension.

6.3.1   Concentric Contractions

The tension in a muscle decreases as it shortens under load. The characteristic curve that describes this effect is called a force-velocity curve, and is seen in Figure 6.10. The usual curve is plotted for a maximum contraction; however, this condition is rarely seen except in athletic events, and then only for short bursts of time. In Figure 6.10 the curves for a 75%, 50%, and 25% contraction are shown as well. Isometric contractions lie along the zero velocity axis of this graph and should be considered as nothing more than a special case within the whole range of possible velocities. It should be noted that this curve represents the characteristics at a certain muscle length. To incorporate length as a variable as well as velocity requires a three-dimensional plot, as will be discussed in Section 6.4.3.

The decrease of tension as the shortening velocity increases has been attributed to two main causes. A major reason appears to be the loss in tension as the cross-bridges in the contractile element break and then reform in a shortened condition. A second cause appears to be the fluid viscosity in both the contractile element and the connective tissue. Such viscosity requires internal force to overcome and therefore results in a lower tendon force. Whichever the cause of loss of tension, it is clear that the total effect is similar to that of viscous friction in a mechanical system,

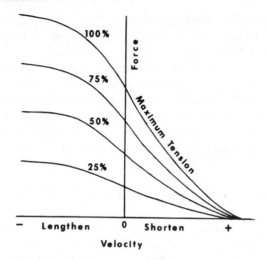

Figure 6.10   Force velocity characteristics of skeletal muscle, showing decrease of tension as muscle shortens and increase as it lengthens.

and can therefore be modeled as some form of fluid damper. More will be said in Section 6.5.

A curve fit of the force-velocity curve was demonstrated by Fenn and March (1935) by the equation:

$$V = V_0 e^{-P/B} - KP \qquad (6.1)$$

where: V is the shortening velocity at any force.
   V_0 is the shortening velocity of the unloaded muscle.
   P is the force.
   B and K are constants.

A few years later Hill (1938) proposed a different mathematical relationship that bore some meaning with regard to the internal thermodynamics. Hill's curve was in the form of a hyperbola:

$$(P+a)(V+b) = (P_0+a)b \qquad (6.2)$$

where: P_0 is the maximum isometric tension.
   a is the coefficient of shortening heat.
   $b = a.V_0/P_0$.
   V_0 is the maximum velocity (when $P = 0$).

More recently this hyperbolic form has been found to fit the empirical constant only during isotonic contractions near resting length.

6.3.2   Eccentric Contractions

The vast majority of research done on isolated muscle during *in vivo* experiments has been concentric contractions. As a result, there is relatively little knowledge about the details of the force-velocity curve as the muscle lengthens. The curve certainly does not follow the detailed mathematical relationships that have been developed for concentric contractions.

This lack of information about eccentric contractions is unfortunate because normal human movement usually involves as much eccentric activity as concentric contractions. If we neglect air and ground friction, level walking involves equal amounts of positive and negative work, and in downhill gait negative work predominates. Figure 6.10 shows the general shape of the force-velocity curve during eccentric contractions. It can be seen that this curve is an extension of the concentric curve.

Experimentally, it is somewhat more difficult to conduct experiments involving eccentric work because an external device must be available to do the work on the human muscle. Such a requirement means that a motor is needed to provide an external force that will always exceed that of the muscle. Experiments on isolated muscle are safe to conduct, but *in vivo* experiments on humans are difficult because such a machine could cause lengthening even past the safe movement of a limb. The excess force could tear the limb apart at the joint! Fool-proof safety mechanisms would have to be installed to prevent such an occurrence.

The reasons given for the forces increasing as the velocity of lengthening increases are similar to those that account for the drop of tension during concentric contractions. First, within the contractile element it is understood that the force required to break the cross-bridge protein links is greater than that required to hold it at its isometric length, and that this force increases as the rate of breaking increases. Second, the viscous friction of shortening is still very much present. However, because the direction of shortening has reversed, the tendon force must now be higher in order to overcome the damping friction. Independent of the exact cause of the force increase with the velocity, it is conceptually safe to model the effect by a viscous damper.

6.3.3   Combination of Length and Velocity on Force

In Section 6.2 and in this section it is evident that the tendon force is a function of both length and velocity. Therefore, a proper representation of both these effects requires a three-dimensional plot like that in Figure

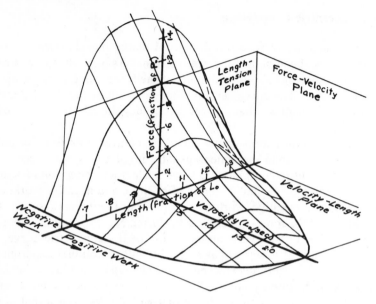

Figure 6.11   Three-dimension plot showing the change is muscle tension as a function of velocity and length. Surface shown is for a maximal contraction.

6.11. The resultant curve is actually a surface, which represents only the maximal contraction condition. The more normal contractions are at a fraction of this maximum, so that surface plots would be required for each level of contraction, say at 75%, 50%, and 25%.

6.3.4   A Typical Time History of a Muscle Contraction

We will use the data provided in Appendices A and B to plot the time course of a typical contraction of the muscle about the ankle. Because the angular changes of the ankle are quite small we can consider that the muscle lengths are proportional to the ankle angle and these length changes are quite small. Thus we can plot the event on a two dimensional force-velocity curve. The lengthening or shortening velocity can be considered proportional to the angular velocity of the ankle and the force is considered proportional to the muscle torque. Thus using *in vivo* data, the traditional force-velocity curve becomes a torque-angular velocity curve. Figure 6.12 is the resultant plot, for the period of time from heel-contact to toe-off. Torques are taken from Table B.1, angular velocities from Table A.11.

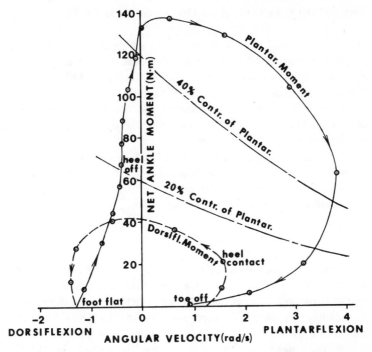

Figure 6.12   Time history of muscle moment—velocity during stance phase of gait. Data was taken from Tables A.5 and B.1. Dorsiflexor moment shown as a dotted line; plantarflexor moment is solid line. Stance begins with negative work, then alternating positive and negative, and finally a major positive work burst later in push-off.

It can be seen from this time course of torque and angular velocity that this common movement is, in fact, quite complex. Contrary to what might be implied by force-velocity curves, a muscle does not operate along any simple curve, but actually goes through a complex combination of force and velocity changes. Also note that the plantarflexor muscles are initially on during dorsiflexion. This situation occurs between the time of heel-contact and flat-foot, when the ground reaction force which acts behind the ankle joint causes the foot to plantarflex. Negative work is being done by the dorsiflexors as they lower the foot to the ground. After flat-foot the plantarflexors dominate to create a torque which tends to slow down the shank as it rotates over the foot which is now fixed on the floor. Again this is negative work. Only after heel-off does positive work begin, as indicated by the simultaneous plantarflexor muscle movement and plantarflexor velocity. This period is the active push-off phase, when most of the "new" energy is put back into the body.

6.4   THE MUSCLE TWITCH AND THE MUSCLE MODEL

Thus far we have said very little about the smallest increment of tension, that of the individual twitch itself. As was described in Section 6.2, each motor unit had its unique time course of tension. Although there are individual differences in each newly recruited motor unit they all have the same characteristic shape. The time-course curve follows quite closely that of the impulse response of a critically damped second order system (Milner-Brown et al., 1973a). The electrical stimulus of a motor unit, as indicated by the motor action potential, is of a short duration and can be considered an impulse. The mechanical response to this impulse is the much longer duration twitch. The general expression for a second-order critically damped impulse response is:

$$F(t) = F_0 \frac{t}{T^2} e^{-t/T} \tag{6.3}$$

The curve is plotted in Figure 6.13 and T is the time for the tension to reach a maximum and F_0 is a constant for that given motor unit. T is the contraction time and is larger for the slow twitch fibers than for the fast twitch motor units, while F_0 increases for the larger fast twitch units. Muscles tested by Buchthal and Schmalbruch (1970) showed a wide range of contraction times. Muscles of the upper limbs had shorter T's compared with the leg muscles. Typical values of T were:

Triceps brachialis	– 44.5 msec	(σ = 9.5 msec)
Tibialis anterior	– 48.0 msec	(σ = 9.0 msec)
Soleus	– 74.0 msec	(σ = 11.0 msec)
Gastrocnemius	– 79.0 msec	(σ = 12.0 msec)

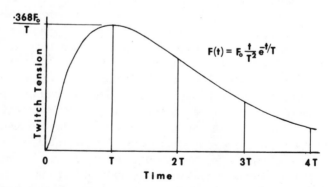

Figure 6.13   Time course of a muscle twitch, modelled as the impulse response of a second order critically-damped system.

These same researchers found that T increased in all muscles as they were cooled; for example, the biceps brachialis had a contraction time that increased from 54 msec at 37°C to 124 msec at 23°C. It is evident that the cause of the delayed buildup of tension is the slower metabolic rates and increased muscle viscosity seen at lower temperatures.

6.4.1   Shape of Graded Contractions

The shape of a voluntary tension curve depends to a certain extent on the shape of the individual muscle twitches. For example, if we elicit a maximal contraction in a certain muscle, the rate of increase of tension depends on the individual motor units and how they are recruited. Even if all motor units were turned on at the same instant and fired at their maximum rate, maximum tension could never be achieved in less than the average contraction time for that muscle. However, as described in Figure 6.2 and depicted in Figure 6.3, the recruitment of motor units does not take place all at once during a voluntary contraction. The smaller slow twitch units are recruited first in accordance with the size principle, with the largest units not being recruited until tension has built up in the smaller units. Thus it can take several hundred milliseconds to reach maximum tension. During voluntary relaxation of the same muscle the drop in tension is governed by the shape of the trailing edge of the twitch curve; this delay in drop of tension is even more pronounced than the rise. This delay combined with the delay in dropping out of the motor units themselves according to the size principle, means that a muscle takes longer to turn off than turn on. Typical turn-on times could be 200 msec and turn-off times 300 msec, as shown in Figure 6.14 where a maximal rapid turn on and off are plotted showing the tension curve and the associated EMG.

6.4.2   Muscle Modeling

A variety of mechanical models of muscle have evolved to describe and predict tension, based on some input stimulation. Crowe (1970) and Gottleib and Agarwal (1971) proposed a contractile component in conjunction with a linear series and parallel elastic component plus a linear viscous damper. Glantz (1974) proposed nonlinear elastic components plus a linear viscous component. Winter (1976) has used a linear mass, spring, and damper system to simulate the second-order, critically damped twitch. The purpose of this section is not to criticize or justify one model versus another, but rather to go over the principles behind the modeling and the components.

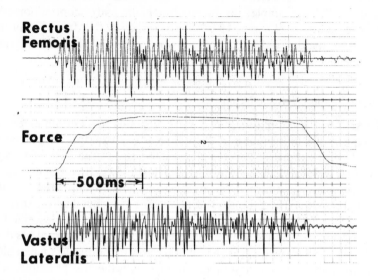

Figure 6.14   Tension buildup and decrease during a rapid maximal voluntary contraction and relaxation. The time to peak tension can be 200 ms or longer, mainly because of the recruitment according to the size principle (see Section 6.2.1) and because of delay between each m.a.p. and twitch tension.

Figure 6.15 shows the force displacement and force-velocity relationships of linear and nonlinear springs and dampers that have been proposed. The symbol for a viscous damper is a "piston in a cylinder," which can be considered full of a fluid of suitable viscosity to represent the constant k. The more common nonlinear models are exponential in form or to a power "a" which is usually greater than 1. This is especially true of viscous friction, which often varies approximately as the velocity squared.

The total model of the passive components can take many configurations as shown in Figure 6.16a. The parallel elastic component can be considered to be in parallel with the damper or a series combination of the damper and series elastic component. For linear components it does not make the slightest difference which configuration is used, because they can be made equivalent. Fung (1971) has shown that:

$$k_1 = k_3 + k_4, \quad \frac{k_1 k_2}{k_1 + k_2} = k_3, \text{ and } \frac{b_1}{k_1 + k_2} = \frac{b_2}{k_4}$$

This means that if either model is known, the other equivalent model can replace it, and it will have the same dynamic characteristics.

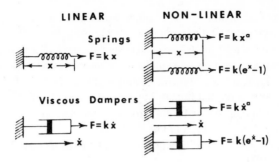

Figure 6.15 Schematic diagram of linear and nonlinear spring and viscous damper elements used to represent passive viscoelastic characteristics of muscle.

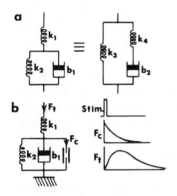

Figure 6.16
a Two equivalent series/parallel arrangement of linear elements of a muscle model.
b Model showing contractile element acting on viscous elastic elements. Twitch tension of the tendon, F_t, results from an exponential tension from the contractile element, F_c.

The total model requires the active contractile component to be represented by some form of force generator. The time-course of the tension from the contractile component is sometimes referred to as its "active state," and this is quite often considered to be an exponential response to a stimulus. Figure 6.16b shows the contractile component combined with the passive components along with the time course of the active state, Fc, and the resultant tendon tension, Ft. Far more advanced models have been proposed, but the mathematical sophistication involved is beyond the scope of this text. Students interested are directed to the references.

REFERENCES

Brandstater, M. E. and Lambert, E. H. "Motor unit anatomy." In *New Developments in Electromyography and Clinical Neurophysiology*, Ed. by J. E. Desmedt, Vol. 1, Karger, Basel, 1973.

Buchthal, F. and Schmalbrugh, H. Contraction times and fibre types in intact human muscle. *Acta Physiol. Scand.* **79**:435–452, 1970.

Crowe, A. A mechanical model of muscle and its application to the intrafusal fibres of mammalian muscle spindle. *J. Biomech.* **3**:583–592, 1970.

Feinstein, B., Lindegard, B., Nyman, E. and Wohlfart, G. Morphological studies of motor units in normal human muscles. *Acta Anat.*, **23**:127–142, 1955.

Fenn, W. O. and Marsh, B. S. Muscular force at different speeds of shortening. *J. Physiol. Lond.* **85**:277–297, 1935.

Fung, Y. C. Comparison of different models of the heart muscle. *J. Biomech.* **4**:289–295, 1971.

Glantz, S. A. A constitutive equation for the passive properties of muscle. *J. Biomech.* **7**:137–145, 1974.

Gordon, A. M., Huxley, A. F. and Julian, F. J. The variation is isometric tension with sarcomere length in vertebrate muscle fibres. *J. Physiol.* **184**:170, 1966.

Gottlieb, G. L. and Agarwal, G. C. Dynamic relationship between isometric muscle tension and the electromyogram in man. *J. Appl. Physiol.* **30**:345–351, 1971.

Henneman, E. "Organization of the spinal chord." In *Medical Physiology*, Ed. by V. B. Mountcastle, 13th ed., Vol. 1, C. V. Mosby, St. Louis, 1974a.

Henneman, E. "Peripheral mechanism involved in the control of muscle." In *Medical Physiology,* Ed. by V. B. Montcastle, 13th ed. Vol. 1, C. V. Mosby, St. Louis, 1974b.

Hill, A. V. The heat of shortening and dynamic constants of muscle. *Proc. Roy. Soc. B.* **126**:136–195, 1938.

Milner-Brown, H. S., Stein, R. B. and Yemm, R. The contractile properties of human motor units during voluntary isometric contractions. *J. Physiol.* **228**:285–306, 1973a.

Milner-Brown, H. S., Stein, R. B. and Yemm, R. The orderly recruitment of human motor units during voluntary isometric contractions. *J. Physiol.* **230**:359–370, 1973b.

Milner-Brown, H. S. and Stein, R. B. The relation between the surface electromyogram and muscular force. *J. Physiol.* **246**:549–569, 1975.

Wani, A. M., Guha, S. K. A model for gradation of tension recruitment and rate coding. *Med. Biol. Eng.* **13**:870–875, 1975.

Wormolts, J. R. and Engel, W. K. "Correlation of motor unit behaviour with histochemical myofiber type in humans by open-biopsy electromyography." In *New Developments in Electromyography and Clinical Neurophysiology,* Ed. by J. E. Desmedt, Vol. 1, Karger, Basel, 1973.

Winter, D. A. Biomechanical model relating EMG to changing isometric tension. *Dig. 11th Internat. Conf. Med. Biol. Eng.* pp 362–363, 1976.

CHAPTER SEVEN

Kinesiological Electromyography

The electrical signal associated with the contraction of a muscle is called an electromyogram or, by its shorthand name, EMG. The study of EMG's, called electromyography, has revealed some basic information; however, much remains to be learned. Voluntary muscular activity results in an EMG that increases in magnitude with the tension. However, there are many variables that can influence the signal at any given time: velocity of shortening or lengthening of the muscle, rate of tension build-up, fatigue, and reflex activity. An understanding of the electrophysiology and the technology of recording is essential to the appreciation of the biomechanical relationships which follow.

7.1 ELECTROPHYSIOLOGY OF MUSCLE CONTRACTION

It is important to realize that muscle tissue conducts electrical potentials somewhat similarly to the way axons transmit action potentials. The name given to this special electrical signal generated in the muscle fibers is muscle action potential (m.a.p.). Electrodes placed on the surface of a muscle or inside the muscle tissue (indwelling electrodes) will record the algebraic sum of all m.a.p.'s being transmitted along the muscle fibers at that point in time. Those motor units far away from the electrode site will result in a smaller m.a.p. than those of similar size near the electrode.

7.1.1 Motor End Plate

For a given muscle there can be a variable number of motor units, each controlled by a motor neuron through special synaptic junctions called motor end plates. An action potential transmitted down the motoneuron (sometimes called the final common pathway) arrives at the motor end plate and triggers a sequence of electrochemical events. A quantum of ACh is released; it crosses the synaptic gap (200 to 500Å wide) and causes a depolarization of the post synaptic membrane. Such a depolarization can be recorded by a suitable microelectrode and is called an end plate

127

potential (EPP). If the EPP is large enough it will reach a critical level and an action potential is initiated in the adjacent muscle fiber membrane.

7.1.2 Sequence of Chemical Events Leading to a Twitch

The beginning of the m.a.p. starts at the Z line of the contractile element (see Section 6.1) by means of an inward spread of the stimulus along the transverse tubular system. This results in a release of Ca^{2+} in the sarcoplasmic reticulum. Ca^{2+} rapidly diffuses to the contractile filaments of actin and myosin, where ATP is hydrolyzed to produce ADP plus heat plus mechanical energy (tension). The mechanical energy manifests itself as an impulsive force at the cross bridges of the contractile element. The time course of the contractile element's force has been the subject of research and has been modeled as described in Section 6.3.2.

7.1.3 Generation of a Muscle Action Potential

The depolarization of the transverse tubular system and the sarcoplasmic reticulum results in a depolarization "wave" along the direction of the muscle fibers. It is this depolarization wavefront and the subsequent repolarization wave that are "seen" by the recording electrodes.

Many types of EMG electrodes have developed over the years, but generally they can be divided into two groups: surface and indwelling (intramuscular). Basmajian (1973) gives a detailed review of the use of different types along with their connectors. Surface electrodes consist of discs of metal, usually silver/silver chloride, of about 1 cm diameter. These electrodes detect the average activity of superficial muscles and give more reproducible results than do indwelling types (Komi and Buskirk, 1973). Smaller discs can be used for smaller muscles. Indwelling electrodes are required, however, for the assessment of fine movements or to record from deep muscles. A needle electrode is nothing more than a fine hypodermic needle with an insulated conductor located inside and bared to the muscle tissue at the open end of the needle; the needle itself forms the other conductor. For research purposes multi-electrode types have been developed to investigate the territory of a motor unit. Fine wire electrodes with about the diameter of human hairs are now widely used. They require a hypodermic needle to insert; after removal of the needle the fine wires with their uninsulated ends remain inside in contact with the muscle tissue.

Indwelling electrodes are influenced not only by waves that actually pass by its conducting surfaces, but also by waves that pass nearby. The same is true of surface electrodes. In Figure 7.1 we see a traveling wave

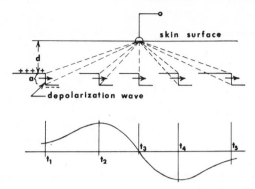

Figure 7.1   Propagation of motor unit action potential wavefront as it passes beneath a recording electrode located on the skin surface. The electrode voltage is a function of the magnitude of the depolarization potential, and the solid angle subtended at the electrode by the wavefront area.

beneath the muscle surface where two electrodes are placed. The potential of any electrode depends on the magnitude of the depolarizing wave, m, the depth of the wave below the skin surface, d, and the area of the leading edge of the wavefront, a (Brown, 1968). The expression for the potential, V, at a point electrode is

$$V = Km\Omega \tag{7.1}$$

where: Ω is the solid angle subtended at the electrode by the wavefront area.

   K is the constant.

   The shape of the waveform produced by the combined depolarization and repolarization is presented in Figure 7.1. It can be seen that the maximum positive voltage is produced when the angle subtended by the wavefront is at a maximum. As the wavefront passes beneath the center of the electrode, the voltage reverses polarity and reaches a maximum in the negative direction, then returns to zero as the wave travels to the end of the muscle. A repolarization wave, if present, would cause the signal to reverse polarity and would produce a positive wave. The depolarization process is quite rapid; thus the leading edge of the wavefront is quite sharp. The repolarization process, on the other hand, is a comparatively slow process and the repolarization wave is not nearly so sharp; therefore, the magnitude of the m.a.p. resulting from the repolarization wave will be quite small. Thus for all practical purposes a biphasic wave will be produced at a single electrode. The effect of a larger surface electrode is

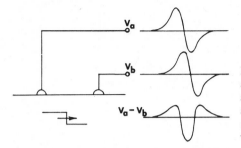

Figure 7.2 Voltage waveform present at two electrodes due to a single propagating wave. Voltage recorded is usually the difference voltage, V_a-V_b, which is triphasic in comparison with biphasic waveform seen at a single electrode.

to widen the first and last waves and to increase the magnitude of the potentials (Brown, 1968).

Most EMG's require two electrodes over the muscle site, so the voltage waveform that is recorded is the difference in potential between the two electrodes. In Figure 7.2 we see that the voltage waveform at each electrode is almost the same, but is shifted in time. Thus a biphasic potential waveform at one electrode usually results in a difference signal which has three phases. The closer the spacing, the more the difference signal looks like a time differentiation of that signal recorded at a single electrode (Kadefors, 1973).

7.1.4   Duration of the Muscle Action Potential

As indicated in the previous section, the larger the surface area the longer the duration of the m.a.p. Thus surface electrodes automatically record longer duration m.a.p.'s than indwelling electrodes (Kadefors, 1973; Basmajian, 1970). However, for a given set of electrodes the duration of the m.a.p.'s is a function of the velocity of the propagating wavefront. The velocity of propagation of the m.a.p. in normals has been found to be about 4 m/sec (Buchthal, Guld, and Rosenfalck, 1955). The faster the velocity the shorter the duration of the m.a.p. Such a relationship has been used to a limited extent in the detection of velocity changes. In fatigue and in certain myopathies (muscle pathologies) the average velocity of the m.a.p.'s that are recruited is reduced; thus the duration of the m.a.p.'s increases (Johansson et al., 1970; Gersten et al., 1965; Kadefors, 1973). During voluntary contractions it has been possible under special laboratory conditions (Milner-Brown and Stein, 1975) to detect the duration and amplitude of muscle action potentials directly from the EMG. However, during unconstrained movements a computer analysis of the total EMG would be necessary to detect a shift in the frequency spectrum (Kwatny et al., 1970), or an autocorrelation analysis can yield the average m.a.p. duration (Person and Mishin, 1964).

7.2   RECORDING OF THE EMG

A biological amplifier of certain specifications is required for the recording of the EMG, whether from surface or indwelling electrodes. It is valuable to discuss the reasons behind these specifications because of the considerable problems in getting a "clean" EMG signal. Such a signal is the summation of m.a.p.'s, and should be undistorted and free of noise or artifacts. Undistorted means that the signal has been amplified linearly over the range of the amplifier and recording system. The larger signals (up to 5 mV) have been amplified as much as the smaller signals (100 μV and below). The most common distortion is overdriving of the amplifier system such that the larger signals appear to be clipped off. Every amplifier has a dynamic range and should be such that the largest EMG signal expected not exceed that range. Noise can be introduced from other sources than the muscle, and can be biological in origin or man-made. An ECG signal picked up by EMG electrodes on the thoracic muscles can be considered unwanted biological noise. Man-made noise usually comes from power lines (hum) or from machinery. Artifacts generally refer to false signals generated by the electrodes themselves or the cabling system. Anyone familiar with EMG recording will recall the lower frequency baseline jumps called movement artifacts, which result from touching of the electrodes and movement of the cables.

The major considerations to be made when specifying the EMG amplifier are:

1. Gain and dynamic range.

2. Input impedance.

3. Frequency response.

4. Common mode rejection.

7.2.1   Amplifier Gain

Surface EMG's have a maximum amplitude of 5 mV peak-to-peak, as recorded during a maximum voluntary contraction. Indwelling electrodes can have a larger amplitude, up to 10 mV. A single m.a.p. has an amplitude of about 100 μV. The noise level of the amplifier is the amplitude of the higher frequency random signal seen when the electrodes are shorted together, and should not exceed 50 μV, preferably 20 μV. The gain of the amplifier is defined as the ratio of the output voltage to the input voltage. For a 2 mV input and a gain of 1000 the output will be 2 V. The exact gain chosen for any given situation will depend on what is to be done with the output signal. The EMG can be recorded on a pen recorder or magnetic tape, viewed on an oscilloscope, or fed straight into a computer. In each case the amplified EMG should not exceed the range of input signals expected by this recording equipment. Fortunately, most of these recording systems have internal amplifiers that can be adjusted to accommodate a wide range of input signals. In general, a good bioamplifier should have a range of gains selectable from 100 to 10,000. Independent of the amplifier gain the amplitude of the signal should be reported as it appears at the electrodes, in millivolts.

7.2.2 Input Impedance

The input impedance (resistance) of a biological amplifier must be sufficiently high so as not to attenuate the EMG signal as it is connected to the input terminals of the amplifier. Consider the amplifier represented in Figure 7.3a. The active input terminals are 1 and 2, with a common terminal c. The need for a three-input terminal amplifier (differential amplifier) will be explained in section 7.2.4.

Each electrode/skin interface has a finite impedance which depends on many factors: thickness of the skin layer, the cleaning of the skin prior to the attachment of the electrodes, the area of the electrode surface, and the temperature of the electrode paste (it warms up from room temperature after attachment). Indwelling electrodes are somewhat special because of the small surface area of bare wire that is in contact with the muscle tissue.

In Figure 7.3b the electrode/skin interface has been replaced with an equivalent resistance. This is a simplification of the actual situation; a correct representation is a more complex impedance to include the capacitance effect between the electrode and the skin. As soon as the amplifier is connected to the electrodes, the minute EMG signal will cause current to flow through the electrode resistances, Rs_1 and Rs_2, to the input impedance of the amplifier, Ri. The current flow through the electrode resistances will cause a voltage drop so that the voltage at the input terminals, V_{in}, will be less than the desired signal V_{emg}. For example, if $Rs_1 = Rs_2 = 10,000$ ohms and $Ri = 80,000$ ohms, a 2 mV EMG signal will give 1.6 mV at V_{in}. A voltage loss of 0.2 mV exists across each of the electrodes. If Rs_1 and Rs_2 were decreased by better skin preparation to 1000 ohms, and Ri were increased to 1 megohm, the 2 mV EMG signal would be reduced only slightly, to 1.998 mV. Thus it is desirable to have input impedances of 1 megohm or higher, and to prepare the skin to reduce the impedance to 1000 ohms or less. For indwelling electrodes the electrode impedance can be as high as 50,000 ohms, so an amplifier with at least 5 megohms input impedance should be used.

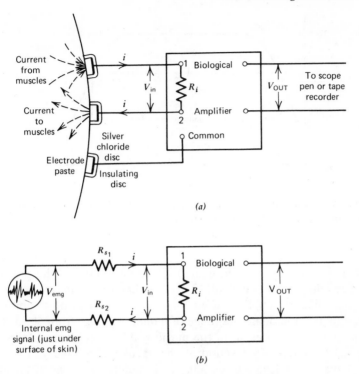

Figure 7.3   Biological amplifier for recording electrode potentials.
a Current resulting from muscle action potentials flows across skin/electrode interface to develop a voltage, V_{in}, at the input terminals of the amplifier. Third, common, electrode is normally required because the amplifier is usually a differential type.
b Equivalent circuit showing electrodes replaced by series resistors, Rs_1 and Rs_2. V_{in} will be nearly equal to V_{emg} if $R_i \gg R_s$.

7.2.3   Frequency Response

The frequency bandwidth of an EMG amplifier should be such as to amplify, without attenuation, all frequencies present in the EMG. The bandwidth of any amplifier, as shown in Figure 7.4, is the difference between the upper cutoff frequency, f_2, and the lower cutoff frequency, f_1. The gain of the amplifier at these cutoff frequencies is .707 of the gain in the mid-frequency region. If we express the mid-frequency gain at 100%, the gain at the cutoff frequencies has dropped to 70.7%, or the power has dropped to $(.707)^2 = .5$. These are also referred to as the half power points. Often amplifier gain is expressed in logarithmic form and expressed in decibels (dB).

$$\text{Gain (dB)} = 20 \log_{10} \text{(linear gain)} \qquad (7.2)$$

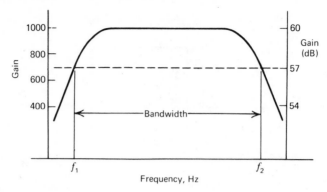

Figure 7.4 Frequency response of the biological amplifier showing a gain of 1000 (60db), and lower and upper cut-off frequencies, f_1 and f_2.

If the linear gain were 1000 the gain in dB would be 60, and the gain at the cutoff frequency would be 57 dB (3 dB less than that at mid-frequency).

In a high fidelity amplifier used for music reproduction f_1 and f_2 are designed to accommodate the range of human hearing, from 50 to 20,000 Hz. All frequencies present in the music will be amplified equally, producing a true undistorted sound at the speakers. Similarly, the EMG should have all its frequencies amplified equally. The spectrum of the EMG has been widely reported in the literature with a range from 5 Hz at the low end to 2000 Hz at the upper end. For surface electrodes the m.a.p.'s are longer in duration and thus don't have much power beyond 1000 Hz. A recommended range for surface EMG is 10 to 1000 Hz, and 20 to 2000 Hz for indwelling electrodes. Figure 7.5 shows a typical EMG spectrum, and it can be seen that most of the signal is concentrated in the band between 20 and 200 Hz with a lesser component extending up to 1000 Hz.

The spectrum of other physiological and noise signals must be considered. ECG signals contain power out to 100 Hz, so it may not be possible to eliminate such interference, especially when monitoring muscle activity around the thoracic region. The major interference comes from power line hum: in North America it is 60 Hz; in Europe it is 50 Hz. Unfortunately hum lies right in the middle of the EMG spectrum, so nothing can be done to filter it out. Movement artifacts, fortunately, lie in the 0 to 10 Hz range and should not cause problems. Unfortunately, some of the lower quality cabling systems can generate large low frequency artifacts that can seriously interfere with the baseline of the EMG recording. Usually such artifacts can be eliminated by good low frequency filtering, by setting f_1 to about 20 Hz. If this fails the only solution is to replace the

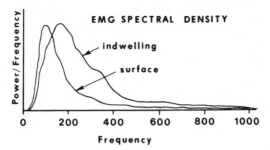

Figure 7.5 Frequency spectrum of EMG as recorded via surface and indwelling electrodes. Higher frequency content of indwelling electrodes is due to closer spacing between electrodes.

cabling or go to the expense of using microamplifiers right at the skin surface.

It is valuable to see the EMG signal as it is filtered using a wide range of bandwidths. Figures 7.6a and b show the results of such filtering.

7.2.4 Common Mode Rejection

The human body is a good conductor, and therefore will act as an antenna to pick up any electromagnetic radiation that is present. The most common radiation comes from domestic power: power cords, fluorescent lighting, and electrical machinery. The resulting interference may be so large as to prevent recording of the EMG. If we were to use an amplifier with a single-ended input we would see the magnitude of this interference. Figure 7.7 depicts hum interference on the active electrode. It appears as a sinusoidal signal, and if the muscle is contracting its EMG would be added. However, hum could be 100 mV, and would drown out even the largest EMG signal.

If we replace the single-ended amplifier with a differential amplifier (Figure 7.8), we can possibly eliminate most of the hum. Such an amplifier takes the difference between the signals on the active terminals. As can be seen this hum interference appears as an equal amplitude on both active terminals. The body acts as an antenna, and all locations pick up the same hum signal. Because this unwanted signal is common to both active terminals it is called a *common* mode signal. At terminal 1 the net signal is $V_{hum} + emg_1$; at terminal 2 it is $V_{hum} + emg_2$. The amplifier has a gain of A, therefore the ideal output signal is:

$$e_o = A \ (e_1 - e_2)$$
$$= A \ (V_{hum} + emg_1 - V_{hum} - emg_2)$$
$$= A \ (emg_1 - emg_2) \tag{7.3}$$

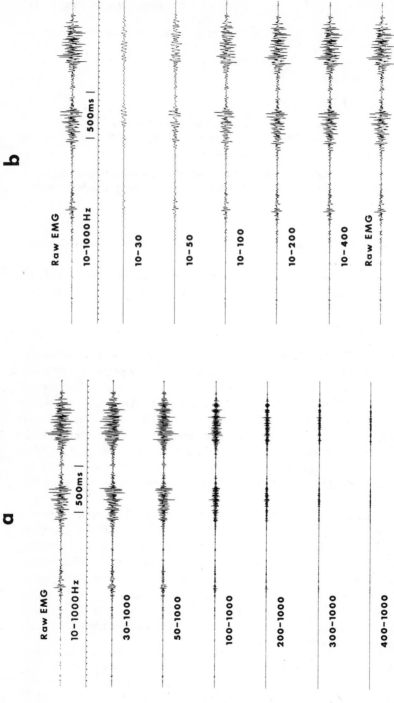

Figure 7.6 Surface EMG filtered with varying cut-off frequencies.
a With lower cut-off frequency varied from 30 to 400 Hz, showing the effect of rejecting lower frequencies.
b With upper cut-off frequency varied from 30 to 400 Hz, showing the loss of higher frequencies.

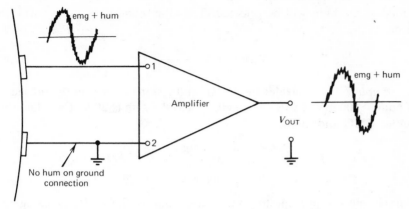

Figure 7.7   A single-ended amplifier showing lack of rejection of hum present on non-grounded active terminal.

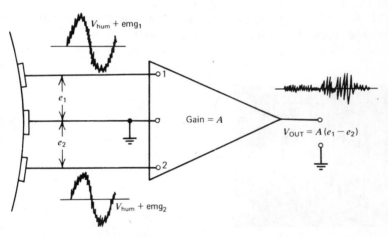

Figure 7.8   Biological amplifier showing how the differential amplifier rejects the common (hum) signal.

The output, e_o, is an amplified version of the difference between the EMG on electrode 1 and 2. No matter how much hum is present at the individual electrodes, it has been removed by a perfect subtraction within the differential amplifier. Unfortunately, a perfect subtraction never occurs, and the measure of how successfully this has been done is given by the common mode rejection ratio (CMRR). If CMRR is 1000:1 then all but

1/1000 of the hum will be rejected. Thus the hum at the output will be given by:

$$V_o(\text{hum}) = \frac{A \times V_{\text{hum}}}{\text{CMRR}} \qquad (7.4)$$

Example 7.1.   Consider the EMG on the skin = 2 mV in the presence of hum = 500 mV. CMRR is 10,000:1 and the gain is 2000. Calculate the output EMG and hum.

$$e_o = A(\text{emg}_1 - \text{emg}_2)$$
$$= 2000 \times 2 \text{ mV} = 4 \text{ V}$$
$$\text{Output hum} = 2000 \times 500 \text{ mV} \div 10,000 = 100 \text{ mV}$$

Hum is always present to some extent unless the EMG is being recorded with battery-powered equipment not in the presence of domestic power. Its magnitude can be seen in the baseline when no EMG is present. Figure 7.9*a* and *b* shows two EMG records, the first with hum quite evident, the latter with negligible hum.

CMRR is often expressed as a logarithmic ratio rather than a linear ratio. The units of this ratio are decibels (dB).

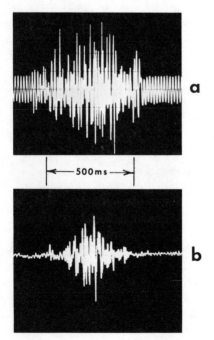

500 ms

a

b

Figure 7.9 Storage scope recordings of an EMG signal.
a With hum present.
b Without hum.

$$\text{CMRR (db)} = 20 \log_{10} \text{CMRR (linear)} \qquad (7.5)$$

If CMRR = 10,000 : 1 then CMRR (dB) = $20 \log_{10} 10{,}000 = 80$ dB. In good quality biological amplifiers the CMRR should be 80 dB or higher.

7.3 PROCESSING OF EMG

Once the EMG signal has been amplified it can be processed for comparison or correlation with other physiological or biomechanical signals. The need for changing the EMG into another processed form is that the raw EMG may not be suitable for recording or correlation. For example, because of the higher frequencies present in the EMG, it is impossible to record it directly on a pen recorder. The low frequency response of most recorders (0 to 60 Hz) means that most of the higher-frequency components of the EMG are not seen.

The more common types of on-line processing are:

(i) Half or full-wave rectification (the latter is also called absolute value).

(ii) Linear envelope detector (half or full-wave rectifier followed by a low pass filter).

(iii) Integration of the full-wave rectified signal over the entire period of the muscle contraction.

(iv) Integration of the full-wave rectified signal for a fixed time, reset to zero, then integration cycle repeated.

(v) Integration of the full-wave rectified signal to a preset level, reset to zero, then integration repeated.

Schematically, these various processing methods are shown in Figure 7.10, and a sample record of a typical EMG processed all five ways is presented and will now be discussed in detail.

7.3.1 Full-Wave Rectification

The full-wave rectifier generates the absolute value of the EMG, usually with a positive polarity. The original raw EMG has a mean value of zero because it is recorded with a biological amplifier with a low-frequency cutoff around 10 Hz. However, the full-wave rectified signal does not cross through zero and therefore has an average or bias level which fluctuates with the strength of the muscle contraction. The quantitative

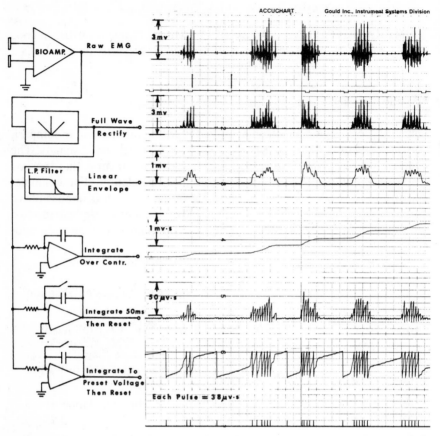

Figure 7.10 Schematic diagram of several common EMG processing systems and the results of simultaneous processing of EMG through these systems. See text for details.

use of the full-wave rectified signal by itself is somewhat limited; it serves as an input to the other processing schemes. The main application of the full-wave rectified signal is in semi-quantitative assessments of the phasic activity of various muscle groups. A visual examination of the amplitude changes of the full-wave rectified signal gives a good indication of the changing contraction level of the muscle. The proper unit for the amplitude of the rectified signal is millivolt, the same as for the original EMG.

7.3.2   Linear Envelope

If we filter the full-wave rectified signal with a lowpass filter, we have what is called the linear envelope. It can best be described as a moving average; it is measured in mV, follows the trend of the EMG, and quite closely resembles the shape of the tension curve. There is considerable confusion concerning the proper name for this signal. Many researchers call this signal an integrated EMG (iEMG); such a term is quite wrong because it can be confused with the mathematical term "integrated," which is a different form of processing.

The main decision to be made with the linear envelope is the choice of the low pass filter. Generally, if the envelope signal is thought to represent the muscle tension the filter should cut off at about 6 Hz and should be at least a second-order type. As was clearly demonstrated in Section 2.6.4, the frequency of human movement lies below 6 Hz, so any signal thought to represent muscle tension should not contain harmonics above 6 Hz. Some of the older electronic equipment used to process EMG's did not have a frequency control, but rather a time-constant setting. Every first-order filter has a time-constant, T, which is related to the cutoff frequency, fc, by the equation:

$$fc = 1/(2\pi T) \qquad (7.6)$$

For a 6-Hz filter a time-constant of about 25 msec applies. If the cutoff frequency is too low, there will be too much smoothing and the envelope will be delayed in following rapid rises and falls in the EMG amplitude. Conversely, if fc is set too high, the envelope will not be delayed in its rise and fall, but it will still have a large noise component in the trend curve.

7.3.3 True Mathematical Integrators

There are several different forms of mathematical integrators. The purpose of an integrator is to measure the "area-under-the-curve." Thus the integration of the full-wave rectified signal will always increase as long as any EMG activity is present. The simplest form starts its integration at some preset time and continues during the time of the muscle activity. At the desired time, which could be a single contraction or a series of contractions, the integrated value can be recorded. The unit of a properly integrated signal is millivolt seconds (mV.s). The only true way to find the average EMG during a given contraction is to divide the integrated value by the time of the contraction; this will yield a value in millivolts.

A second form of integrator involves a resetting of the integrated signal to zero at regular intervals of time (40 to 200 msec). Such a scheme yields a series of peaks which represent the trend of the EMG amplitude with time. Each peak represents the average EMG over the previous time interval, and the series of peaks is a "moving" average. Each peak has units of mV.s so that the sum of all the peaks during a given contraction yields the total integrated signal, as described in the previous paragraph.

There is a close similarity in this reset, integrated curve and the linear envelope. Both follow the trend of muscle activity. If the rest time is too high it will not be able to follow rapid fluctuations of EMG activity, and if it is reset too frequently there will be noise present in the trendline. If the integrated peaks are divided by the integration time the amplitude of the signal can again be reported in millivolts.

A third common form of integration uses a voltage level reset. The integration begins before the contraction. If the muscle activity is high, the integrator will rapidly charge up to the reset level; if the activity is low, it will take longer to reach reset. Thus the strength of the muscle contraction is measured by the frequency of the resets. High frequency of reset pulses is high muscle activity; low frequency is low muscle activity. Intuitively such a relationship is attractive to neurophysiologists because it has a similarity to the action potential rate present in the neural system. The total number of counts over a given time is proportional to the EMG activity. Thus if the threshold voltage level and the gain of the integrator are known, the total EMG activity (mV.s) can be determined.

7.4   RELATIONSHIP BETWEEN EMG AND BIOMECHANICAL VARIABLES

The major reason for processing the basic EMG is to derive a relationship between it and some measure of muscle function. A question that has been posed for years is "How valuable is the EMG in predicting muscle tension?" Such a relationship is very attractive because it would give an inexpensive and noninvasive way of monitoring muscle tension. Also, there may be information in the EMG concerning muscle metabolism, power, fatigue state, or contractile elements recruited.

7.4.1   EMG versus Isometric Tension

Bouisset (1973) has presented an excellent review of the state of knowledge regarding the EMG and muscle tension in normal isometric contractions. The EMG processed through a linear envelope detector has been widely used to compare the EMG/tension relationship, especially if the tension is changing with time. If constant tension experiments are done it is sufficient to calculate the average of the full-wave rectified signal, which is the same as that derived from a long time-constant linear envelope circuit. Both linear and nonlinear relationships between EMG amplitude and tension have been discovered. Typical of the work reporting linear relationships is an early study by Lippold (1952) on the calf muscles of man. Vredenbregt and Rau (1973), on the other hand, found quite

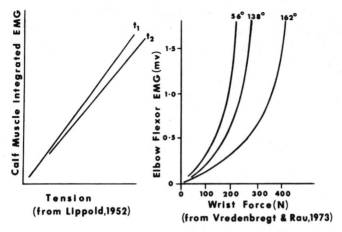

Figure 7.11   Relationship between amplitude of EMG and the tension in the muscle in isometric contraction.

nonlinear relationships between tension and EMG in the elbow flexors over a wide range of joint angles. Both these studies were in effect static calibrations of the muscle under certain length conditions; a reproduction of their results is presented in Figure 7.11.

Another way of representing the level of EMG activity is to count the action potentials over a given period of time. Close and colleagues (1960) showed a linear relationship between the count rate and the integrated EMG, so it was not surprising that the count rate increased with muscle tension in an almost linear fashion.

The relationship between force and linear envelope EMG also holds during dynamic changes of tension. Inman and colleagues (1952) first demonstrated this by a series of force transducer signals which were closely matched by the envelope EMG from the muscle generating the isometric force. Gottlieb and Agarwal (1971) mathematically modelled this relationship with a second order low-pass system (e.g., a low-pass filter). Under dynamic contraction conditions, the tension is seen to lag behind the EMG signal, as shown in Figure 7.12a and b. The delay is due to the fact that the twitch corresponding to each m.a.p. reaches its peak from 40 to 100 msec afterward. Thus, as each motor unit is recruited, the resulting summation of twitch forces will also have a similar delay behind the interference pattern we call the EMG.

In spite of the reasonably reproducible relationships, the question still arises as to how valid these relationships are for dynamic conditions when many muscles act across the same joint:

1. How does the relationship change with length? Does the length change merely alter the mechanical advantage of the muscle, or does the changing overlap of the muscle fibers affect the EMG itself (Vredenbregt and Rau, 1973)?

2. How do other agonist muscles share the load at that joint, especially if some of the muscles have more than one function (Vredenbregt and Rau, 1973)?

3. In many movements there is antagonist activity. How much does this alter the force being predicted by creating an extra unknown force?

With the present state of knowledge it appears that a suitably calibrated linear envelope EMG can be used as a coarse predictor of muscle tension for muscles whose length is not changing rapidly.

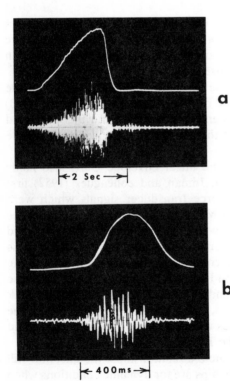

Figure 7.12 EMG and muscle tension recorded on a storage scope during varying isometric contractions of the biceps muscles.
a During a gradual buildup and rapid relaxation.
b During a short 400 ms contraction. Note the delay between the EMG and the initial build-up of tension, time to reach maximum tension, and drop of tension after the EMG has ceased.

7.4.2   EMG During Muscle Shortening and Lengthening

In order for a muscle to do positive or negative work it must also undergo length changes while it is creating tension. Thus it is important to see how well the EMG can predict tension under these more realistic conditions. The major study to date has been reported by Komi (1973); in it, the subject did both positive and negative work on an isokinetic muscle-testing machine. The subject was asked to generate maximal tension while the muscle lengthened or shortened at controlled velocities. The basic finding was that the EMG amplitude remained fairly constant in spite of decreased tension during shortening (Section 6.4.1), and increased tension during eccentric contractions. Such results support the theory that the EMG amplitude indicates the state of tension of the contractile element, which is quite different from the tension recorded at the tendon (Section 6.5.2). Also, these results indicate that the EMG amplitude to do negative work is less than that required to do the same amount of positive work. If the EMG amplitude is a relative measure of muscle metabolism, such a finding supports the experiments that found negative work to have somewhat less metabolic cost than positive work.

7.4.3   EMG Changes During Fatigue

Muscle fatigue occurs when the muscle tissue cannot supply the metabolism at the contractile element, either because of ischemia (insufficient oxygen) or local depletion of any of the metabolic substrates. Mechanically, fatigue manifests itself by decreased tension, assuming that the muscle activity remains constant, as indicated by a constant EMG or stimulation rate. Conversely, to maintain a constant tension after onset of fatigue requires increased motor unit recruitment (Vredenbregt and Rau, 1973). Such findings indicate that all or some of the motor units are decreasing their twitch tensions and must be reinforced by newly recruited units.

Fatigue not only reduces the twitch force but may also alter the shape of its motor action potential. It is not possible to see the changes of shape of the individual m.a.p.'s in a heavy voluntary contraction. However, an autocorrelation will show the shifts in the average duration of the recruited m.a.p. (Section 7.1.4). Also, the EMG spectrum may shift to reflect these duration changes; Kadefors, et al. (1973) found that the higher-frequency components decreased. Such an increase in the m.a.p. duration can be due to two possible causes:

(i) Lower conduction velocity of some or all action potentials (Mortimer, 1970).

(ii) The slower motor units with their longer duration m.a.p.'s remain active while some of the faster motor units with shorter m.a.p.'s have dropped out.

A third change in the EMG associated with fatigue is a tendency for the motor units to fire more synchronously. Normally, each motor unit fires quite independently of the others in the same muscle, so that the EMG can be considered as an "interference" pattern consisting of the summation of a number of randomly spaced m.a.p.'s. However, during fatigue a tremor is evident and is seen in both the tension tracings and the EMG. During a voluntary isometric contraction the tension tracing has fluctuations around 10 Hz. The cause of the fluctuations is neurological, as evidenced by similar fluctuations in the EMG amplitude. Such fluctuations can only be caused by a change in the firing pattern of the motor units such that they tend to fire in synchronized bursts.

REFERENCES

Basmajian, J. V. Electrodes and electrode connectors. In *New Developments in Electromyography and Clinical Neurophysiology*, Ed. by J. E. Desmedt, Vol. 1, pp. 502–510 (Karger, Basel 1973).

Bigland, B. and Lippold, O. C. J. The relation between force, velocity and integrated electrical activity in human muscles. *J. Physiol. Lond.* 123:214–224, 1954.

Brown, B. H. Theoretical and experimental waveform analysis of human compound nerve action potentials using surface electrodes. *Med. Biol. Eng.* 6:375–386, 1968.

Bouisset, S. EMG and muscle force in normal motor activities. In *New Developments in Electromyography and Clinical Neurophysiology*, Ed. by J. E. Desmedt, Vol. 1, pp. 547–583 (Karger, Basel, 1973).

Boyd, D. C., Lawrence, P. D., and Bratty, P. J. A. On modelling the single motor unit action potential. *IEEE Trans. Biomed. Eng.* BME-25:236–242, 1978.

Buchthal, F., Guld, C., and Rosenfalck, P. Propagation velocity in electrically activated fibers in man. *Acta Physiol. Scand.* 34:75–89, 1955.

Buller, A. J. and Lewis, D. M. Rate of tension developed in isometric tetanic contraction of mammalian skeletal muscle. *J. Physiol.* 176:337–354, 1965.

Close, J. R., Nickel, E. D. and Todd, A. B. Motor unit action potential counts. *J. Bone Jt. Surg.* 42-A. 1207–1222, 1960.

Gottlieb, G. L. and Agarwal, G. C. Filtering of electromyographic signals. *Amer. J. Phys. Med.* **49**:142–146, 1970.

Gottlieb, G. L. and Agarwal, G. C. Dynamic relationship between isometric muscle tension and the electromyogram in man. *J. Appl. Physiol.* **30**:345–351, 1971.

Gusten, J. W., Cenkovich, F. S., Jones, G. D. Harmonic analysis of normal and abnormal electromyograms. *Amer. J. Phys. Med.* **44**:235–240, 1965.

Inman, V. T., Ralston, H. J. Saunders, J. B., Feinstein, B. and Wright, E. W. Relation of human electromyogram to muscular tension. *Electroencephalogr. Clin. Neurophysiol.* **4**:187–194, 1952.

Johansson, S., Larsson, L. E. and Ortengren, R. An automated method for the frequency analysis of myoelectric signals evaluated by an investigation of the spectral changes following strong sustained contractions. *Med. Biol. Eng.* **8**:257–264, 1970.

Kadefors, R. Myo-electric signal processing as an estimation problem. In *New Developments in Electromyography and Clinical Neurophysiology*, Ed. by J. E. Desmedt, Vol. 1, pp. 519–532 (Karger, Basel, 1973).

Kadefors, R., Petersen, I. and Broman, H. Spectral analysis of events in the electromyogram. In *New Developments in Electromyography and Clinical Neurophysiology*, Ed. by J. E. Desmedt, Vol. 1, pp. 628–637 (Karger, Basel, 1973).

Komi, P.V. and Buskirk, E. R. Reproducibility of electromyographic measures with inserted wire electrodes and surface electrodes. *Electromyography* **10**:357–367, 1970.

Komi, P. V. Relationship between muscle tension, EMG, and velocity of contraction under concentric and eccentric work. In *New Developments in Electromyography and Clinical Neurophysiology*, Ed. by J. E. Desmedt, Vol. 1, pp. 596–606 (Karger, Basel, 1973).

Kwatny, E., Thomas, D. H. and Kwatny, H. G. An application of signal processing techniques to the study of myoelectric signals. *IEEE Trans. Biomed. Eng.* **BME-17**:303–312, 1970.

Lippold, O. C. J. The relationship between integrated action potentials in a human muscle and its isometric tension. *J. Physiol. Lond.* **177**:492–499, 1952.

Milner-Brown, H. S. and Stein, R. B. The relation between surface electromyogram and muscular force. *J. Physiol.* **246**:549–569, 1975.

Mortimer, J. T., Magnusson, R. and Petersen, I. Conduction velocity in ischemic muscle: Effect on EMG frequency spectrum. Amer. J. Physiol. **219**;1324–1329, 1970.

Nelson, A. J., Moffroid, M. and Whipple, R. Relationship of integrated electromyographic discharge to isokinetic contractions. In *New Developments in Electromyography and Clinical Neurophysiology*,. Ed. by J. E. Desmedt, Vol. 1, pp. 584–595 (Karger, Basel, 1973).

Person, R. S. and Mishin, L. N. Auto and crosscorrelation analysis of the electrical activity of muscles. *Med. Biol. Eng.* **2**:155–159, 1964.

Vredenbregt, J. and Rau, G. Surface electromyography in relation to force, muscle length and endurance. In *New Developments in Electromyography and Clinical Neurophysiology*, Ed. by J. E. Desmedt, Vol. 1, pp. 607–622 (Karger, Basel, 1973).

Zuniga, E. N. and Simons, D. G. Non-linear relationship between averaged electromyogram potential and muscle tension in normal subjects. *Arch. Phys. Med.* **50**:613–620, 1969.

APPENDIX A
Anthropometric, Kinematic, and Force Plate Data

Table A.1 Anatomical Location of Markets, Unit, and Timing Information

LOCOMOTION LABORATORY - DEPARTMENT OF KINESIOLOGY

CODE NO.:_____ SUBJECT/PATIENT ANTHROPOMETRIC DATA SHEET DATE:_____

NAME_____ HEIGHT:_____WEIGHT: 80Kg AGE:____SEX:___

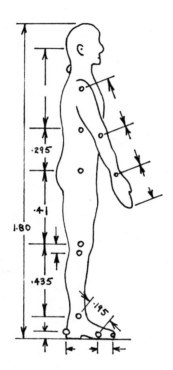

Anatomical Location of Markers

Rib — midline of rib cage
       half way between
       illiac crest and
       shoulder

Hip — greater trochanter

Knee — lateral femoral
       epicondyle (about
       2 cm. above knee line)

Ankle — lateral malleolus of
        fibula

Heel — about 2 cm above
       ground in line with
       rear of shoe

Meta — 5th metatarsal
       phalangeal joint

Toe — about 2 cm. above
       sole in line with
       front of shoe

Units & Timing

1. All raw and filtered coordinates are in cm.

2. All kinematic and kinetic variables are in units indicated.

3. Frame number 1 is at right heel contact, prior frames are labelled -3,-2, -1,0 and are "lost" due to digital filtering and acceleration calculations.

4. Time between frames = 1/60 s.

5. Labelling of frames is approximate, as determined by visual inspection during conversion of film coordinates.

    HSR — heel strike right   TOR — toe off right
    HSL — heel strike left   TOL — toe off left

Table A.2 Anthropometric Data

SEGMENT	DEFINITION	SEG WEIGHT/ TOT BODY WT	CENTER OF MASS/ SEGMENT LENGTH		RADIUS OF GYRATION/ SEGMENT LENGTH			DENSITY
			PROX	DIST	C of G	PROX	DIST	
Hand	Wrist Axis/Knuckle II Middle Finger	.006 M	.506	.494 P	.297	.587	.577 M	1.16
Forearm	Elbow Axis/Ulnar Styloid	.016 M	.430	.570 P	.303	.526	.647 M	1.13
Upper Arm	Glenohumeral Axis/Elbow Axis	.028 M	.436	.564 P	.322	.542	.645 M	1.07
Forearm & Hand	Elbow Axis/Ulnar Styloid	.022 M	.682	.318 P	.468	.827	.565 P	1.14
Total Arm	Glenohumeral Joint/Ulnar Styloid	.050 M	.530	.470 P	.368	.645	.596 P	1.11
Foot	Lateral Malleolus/Head Metatarsal II	.0145 M	.50	.50 P	.475	.690	.690 P	1.10
Shank	Femoral Condyles/Medial Malleolus	.0465 M	.433	.567 P	.302	.528	.643 M	1.09
Thigh	Greater Trochanter/Femoral Condyles	.100 M	.433	.567 P	.323	.540	.653 M	1.05
Foot & Shank	Femoral Condyles/Medial Malleolus	.061 M	.606	.394 P	.416	.735	.572 P	1.09
Total Leg	Greater Trochanter/Medial Malleolus	.161 M	.447	.553 P	.326	.560	.650 P	1.06

151

Table A.2 (Continued)

Head & Neck	C7-T1 & 1st Rib/Ear Canal	.081 M	1.000	– PC	.495	1.116	– PC	1.11
Shoulder Mass	Sternoclavicular Joint/Glenohumeral Axis		.712	.288				1.04
Thorax	C7-T1/T12-L1 & Diaphragm*	.216 PC	.82	.18				0.92
Abdomen	T12-L1/L4-L5*	.139 LC	.44	.56				
Pelvis	L4-L5/Greater Trochanter*	.142 LC	.105	.895				
Thorax & Abdomen	C7-T1/L4-L5*	.355 LC	.63	.37				
Abdomen & Pelvis	T12-L1/Greater Trochanter*	.281 PC	.27	.73				1.01
Trunk	Greater Trochanter/Glenohumeral Joint*	.497 M	.50	.50				1.03
Trunk Head Neck	Greater Trochanter/Glenohumeral Joint*	.578 MC	.66	.34 P	.503	.830	.607 M	
H.A.T.	Greater Trochanter/Glenohumeral Joint*	.678 MC	.626	.374 PC	.496	.798	.621 PC	
H.A.T.	Greater Trochanter/Mid Rib	.678	1.142		.903	1.456		

* NOTE: These segments are presented relative to the length between the Greater Trochanter and the Glenohumeral Joint.

SOURCE CODES
    M – Dempster via Miller & Nelson
    P – Dempster via Plagenhoef
    L – Dempster via Plagenhoef from living subjects
    C – Calculated

Table A.3 Raw Input Coordinate Data

RAW INPUT DATA LISTING:

FRAME	BASE RIB		R. HIP		R. KNEE		R. FIBUL		R. ANKLE		R. HEEL		R. META		R. TOE	
	X	Y	X	Y	X	Y	X	Y	X	Y	X	Y	X	Y	X	Y
-3	32.0	121.2	34.7	92.4	51.4	55.5	51.3	50.7	67.8	15.0	66.2	3.8	87.4	12.3	91.5	16.1
-2	34.6	121.0	37.8	92.1	54.4	55.2	54.2	50.3	71.4	15.2	70.0	3.7	91.1	13.1	95.0	17.4
-1	37.1	120.4	40.6	91.3	57.1	54.8	56.9	49.8	74.1	14.8	72.9	3.3	93.7	13.1	97.6	17.2
0	39.9	120.3	43.5	91.3	60.2	54.8	59.9	49.9	76.3	14.1	75.0	2.6	95.8	12.4	99.6	16.8
1	42.8	119.9	46.7	90.6	63.5	54.1	63.3	49.3	77.8	13.0	76.4	1.8	97.4	10.7	101.3	15.3
2	45.5	120.0	49.6	90.8	66.3	54.4	66.0	49.3	78.7	12.3	76.4	2.1	98.1	9.3	102.1	13.4
3	48.9	120.0	53.1	90.6	69.6	53.9	69.1	49.2	80.7	11.9	76.9	2.2	99.3	7.2	103.8	10.7
4	51.7	119.9	56.1	90.8	72.3	54.4	71.9	49.5	82.1	11.8	77.4	2.6	100.1	5.1	104.9	8.3
5	55.2	119.5	59.6	91.2	75.7	54.4	75.3	49.9	83.2	11.2	77.6	2.3	100.7	3.6	105.4	6.4
6	57.9	119.6	62.4	91.6	78.8	54.9	78.1	50.2	83.8	11.2	77.9	2.5	101.0	3.0	105.9	5.3
7	60.6	120.0	64.9	92.1	81.3	55.2	80.7	50.2	84.4	11.3	78.2	2.6	101.2	2.1	106.1	4.1
8	62.9	120.4	67.4	92.3	83.6	55.2	82.6	50.6	84.8	11.1	78.2	2.6	101.2	1.6	106.1	3.7
9	65.2	120.6	69.6	92.3	85.2	54.9	84.1	50.0	84.6	10.9	78.1	2.3	101.0	1.6	106.1	3.1
10	67.6	121.0	71.9	92.8	86.6	54.8	85.3	50.3	84.9	11.1	77.8	2.3	101.2	1.5	106.1	3.1
11	69.8	121.8	74.2	93.1	87.6	54.8	86.4	50.2	84.7	11.1	77.9	2.5	101.0	1.5	106.1	3.1
12	72.4	121.8	76.5	93.6	88.5	54.5	87.1	50.3	84.7	10.9	77.8	2.6	100.8	1.2	105.9	2.9
13	74.6	122.5	78.4	94.0	89.4	54.7	88.0	50.4	84.7	11.2	78.0	2.5	100.7	1.5	106.1	2.9
14	77.1	122.9	80.8	94.6	90.4	54.4	88.7	50.2	84.9	10.9	77.8	2.3	100.7	1.1	106.2	2.9
15	79.7	123.3	82.8	94.6	90.9	54.4	89.0	49.8	85.0	11.2	78.1	2.5	100.7	1.1	105.9	2.3
16	82.0	124.0	84.7	95.0	91.8	54.5	89.8	50.2	85.2	11.2	78.0	2.6	100.8	1.4	106.3	2.6
17	84.1	124.5	86.5	95.1	92.3	54.7	90.1	50.2	85.4	11.2	78.0	2.7	100.8	0.9	106.3	2.6
18	86.6	124.9	88.4	95.7	92.8	54.7	90.9	50.3	85.4	11.3	77.9	2.5	101.1	1.4	106.1	2.6
19	88.9	124.8	90.0	95.3	93.5	54.0	91.1	50.2	85.0	11.2	77.9	2.5	100.9	1.1	106.1	2.3
20	91.0	124.5	91.7	95.4	93.8	54.0	91.6	50.0	85.4	11.5	78.4	2.7	100.9	0.9	106.6	2.2
21	93.1	124.7	93.7	95.1	94.6	54.1	92.3	50.0	85.3	11.5	78.6	2.7	101.2	0.9	106.7	2.5
22	95.5	124.4	95.5	95.1	95.6	54.1	93.0	49.9	86.0	11.5	78.6	3.0	101.2	0.9	106.8	2.5
23	97.9	124.4	97.4	95.1	96.3	54.4	93.5	50.3	85.8	11.8	78.4	3.3	100.9	1.1	106.6	2.7
24	99.6	124.3	98.9	95.1	97.7	54.3	94.1	50.2	86.1	11.8	78.7	3.1	101.1	1.1	106.8	2.7
25	101.8	123.8	100.9	94.9	98.4	54.3	94.7	50.0	86.1	11.9	78.7	3.4	101.3	0.8	106.8	2.3
26	103.9	123.8	102.8	94.9	99.6	54.1	95.4	50.4	86.6	12.2	78.9	3.4	100.9	1.2	106.8	2.3
27	106.3	123.2	104.8	94.6	100.1	54.4	96.4	50.3	86.3	12.3	78.5	4.1	101.2	0.7	107.0	2.2
28	108.3	123.4	106.6	94.3	101.5	54.4	96.8	50.6	86.4	12.6	78.9	4.8	100.8	0.9	107.0	2.3
29	110.6	122.9	109.0	94.6			97.9	50.7					100.9	1.1	106.7	2.2

Table A.3 *(Continued)*

FRAME	BASE RIB X	Y	R. HIP X	Y	R. KNEE X	Y	R. FIBUL X	Y	R. ANKLE X	Y	R. HEEL X	Y	R. META X	Y	R. TOE X	Y
30	113.3	122.5	111.2	94.2	102.8	54.3	99.2	50.6	86.9	13.1	79.0	5.1	101.0	0.8	106.7	2.1
31	115.6	122.2	113.1	93.8	104.1	54.5	100.6	50.9	87.3	13.5	79.1	5.5	100.7	0.9	106.9	2.2
32	118.1	121.9	115.6	93.3	105.8	54.5	102.4	50.9	87.6	14.2	79.3	6.4	100.7	0.7	106.9	2.1
33	120.1	121.5	117.8	93.8	107.6	54.5	104.0	51.3	88.1	14.8	79.5	7.1	100.8	0.8	106.9	2.2
34	122.5	121.1	119.9	93.2	109.6	54.4	105.6	50.9	88.5	15.3	79.6	8.2	100.8	1.2	106.7	2.2
35	125.1	121.0	122.7	93.2	112.2	54.4	108.2	50.9	89.9	16.0	80.5	9.6	101.5	1.1	107.1	2.3
36	127.3	121.0	125.1	92.7	114.7	54.3	110.9	51.1	90.8	16.5	81.1	10.0	101.7	1.1	107.3	2.3
37	130.7	120.3	128.2	92.0	117.8	53.6	114.0	50.6	92.4	17.5	82.6	12.0	101.5	1.1	107.2	2.2
38	133.2	120.0	130.7	92.0	121.3	53.3	117.2	50.3	93.7	18.7	83.1	13.9	101.5	0.9	107.2	2.2
39	136.1	120.1	133.6	92.3	124.7	53.3	120.6	50.3	95.2	19.4	84.8	16.3	101.4	0.9	107.2	2.3
40	138.6	120.0	136.6	91.4	128.4	52.4	124.7	49.8	97.6	20.5	86.5	18.3	101.4	0.8	107.0	1.9
41	140.8	119.7	139.4	91.3	132.6	52.1	128.8	49.1	100.2	21.6	89.1	21.1	101.4	1.8	107.9	1.5
42	143.6	120.3	142.5	91.3	137.4	51.5	133.7	49.1	103.4	23.2	92.3	23.9	102.2	2.7	107.9	1.4
43	146.5	120.3	145.5	91.3	142.9	51.4	138.7	49.2	107.4	24.7	96.4	26.1	104.4	4.2	108.7	1.2
44	147.9	120.3	147.7	91.2	146.9	51.0	143.1	48.7	110.8	25.8	99.9	27.5	106.9	5.2	111.0	1.6
45	150.2	120.4	150.5	91.2	151.4	51.1	148.0	48.8	115.4	26.6	104.1	28.2	111.7	5.9	115.4	2.3
46	152.4	121.0	153.3	91.3	155.9	51.7	152.4	49.8	119.4	27.2	108.2	28.0	117.2	6.4	121.4	3.1
47	154.2	121.2	155.3	92.0	160.4	52.5	156.7	49.8	124.1	27.3	112.7	27.2	123.5	6.0	127.4	3.4
48	156.6	121.6	158.1	92.3	164.8	52.8	161.1	50.0	129.5	26.8	117.9	26.2	129.8	6.0	134.4	3.0
49	158.8	122.3	160.6	93.1	168.5	54.1	165.1	51.3	134.2	26.4	122.8	24.6	136.6	5.9	141.3	3.1
50	160.8	122.9	162.8	93.6	172.3	54.5	169.0	51.5	139.5	24.7	128.3	22.6	143.6	5.2	148.5	2.9
51	162.9	123.6	165.3	93.3	175.3	55.4	172.9	52.2	145.0	23.8	134.2	22.6	151.1	4.3	156.4	2.2
52	164.9	123.7	167.8	94.3	179.0	55.9	176.4	52.6	150.0	21.6	140.0	19.1	158.8	3.3	164.2	2.1
53	167.4	124.1	170.4	94.4	183.3	56.4	180.4	53.4	157.4	20.2	147.2	14.5	166.2	2.9	172.5	1.9
54	169.3	124.5	172.3	95.1	185.8	57.3	183.2	54.0	162.8	18.9	153.4	11.8	174.2	2.3	179.8	1.9
55	171.4	124.5	174.7	95.1	189.2	57.7	186.5	54.1	169.5	17.2	160.2	9.6	182.4	1.9	187.7	1.8
56	173.7	124.4	177.1	95.1	191.8	57.7	189.3	53.9	175.6	15.7	167.3	7.1	189.6	1.9	195.4	2.5
57	176.0	124.4	179.4	95.1	194.4	57.7	192.2	53.9	182.4	14.5	174.2	5.2	197.0	2.1	202.4	2.9
58	178.2	124.4	181.1	95.0	196.8	57.8	195.1	53.9	188.9	13.3	181.6	4.1	204.6	2.1	210.0	3.6
59	180.4	124.7	183.7	95.1	199.5	57.7	197.3	53.9	195.1	13.3	189.1	3.7	212.1	2.5	217.3	4.9
60	182.7	124.1	185.8	95.5	201.7	57.4	200.2	53.3	201.5	13.0	195.5	2.7	218.6	2.6	224.5	4.9
61	185.2	124.0	188.1	94.3	203.9	57.1	202.5	52.8	207.3	12.7	202.5	2.6	225.2	3.6	230.8	6.1
62	187.2	124.0	189.7	94.3	205.8	56.6	204.7	52.1	212.9	12.7	208.4	2.2	231.0	5.1	236.4	7.8
63	189.5	123.2	192.1	94.3	207.9	56.2	207.1	51.5	218.2	12.7	214.4	2.2	237.0	6.8	242.0	10.0
64	191.5	122.8	194.2	93.4	210.1	55.9	209.3	50.9	222.5	13.7	219.8	2.2	242.2	8.5	246.5	12.0
65	194.2	122.3	196.8	92.9	212.5	55.2	212.1	50.6	226.7	13.7	224.6	2.5	246.8	10.1	250.6	13.9
66	196.5	121.8	199.5	92.5	214.9	55.1	214.7	50.6	230.2	13.9	228.6	3.0	250.2	11.5	254.3	15.6
67	198.7	121.4	202.1	92.5	217.5	55.1	217.3	50.2	233.0	13.8	231.1	2.7	252.7	11.6	256.6	15.0
68	201.2	121.4	204.4	92.5	220.4	54.8	220.2	50.0	235.1	13.8	233.5	2.1	254.8	10.8	258.6	15.0
69	203.9	121.2	207.7	92.0	223.7	54.7	223.0	49.8	237.1	12.7	234.9	1.5	256.5	9.7	260.4	14.2
70	206.2	121.2	210.1	91.2	226.1	54.7	225.7	49.3	238.0	12.4	235.3	1.5	257.5	8.3	261.4	12.6
71	208.7	120.7	212.9	91.2	228.8	54.3	228.2	49.3	238.7	11.2	235.3	1.2	257.6	7.1	261.8	10.5

Table A.3 (Continued)

FRAME	BASE RIB X	Y	R. HIP X	Y	R. KNEE X	Y	R. FIBUL X	Y	R. ANKLE X	Y	R. HEEL X	Y	R. META X	Y	R. TOE X	Y
72	211.8	120.7	215.9	91.3	231.5	54.0	231.0	49.2	240.3	11.1	235.8	0.9	258.6	5.3	263.2	8.5
73	214.3	120.3	218.7	91.3	234.0	53.9	233.3	49.2	241.5	10.7	236.2	1.4	259.1	3.6	263.3	6.3
74	217.3	120.8	221.6	92.5	236.7	54.5	236.0	49.8	241.8	10.8	236.7	1.4	259.4	2.9	263.8	5.5
75	219.5	121.0	223.9	92.0	239.1	54.4	239.2	49.8	242.2	10.4	235.9	1.4	259.3	1.6	263.9	3.8
76	222.3	120.4	226.5	92.3	241.7	54.7	240.8	49.8	243.5	10.3	236.2	1.5	259.7	1.1	264.2	3.0
77	225.0	120.8	229.1	92.5	243.7	54.5	242.1	49.8	243.7	10.0	236.6	1.6	259.9	0.9	264.4	2.9
78	227.2	121.4	231.5	93.1	245.0	54.5	244.1	49.9	243.3	10.0	236.6	1.5	259.8	0.8	264.5	2.6
79	229.2	121.5	233.6	93.1	246.1	54.1	244.7	49.9	243.8	10.0	236.4	1.5	259.8	0.8	264.4	2.2
80	231.8	122.1	236.0	93.4	247.8	54.1	247.3	49.5	243.8	10.5	236.8	1.8	260.4	0.9	264.9	2.2
81	234.4	122.6	237.8	93.5	248.5	54.1	244.5	49.5	243.9	9.7	236.8	1.8	260.0	0.9	264.8	2.3
82	237.0	122.9	240.2	93.8	250.0	54.0	248.2	49.3	243.9	10.4	236.8	1.8	260.2	0.9	265.2	2.2
83	239.4	122.9	241.9	94.3	250.7	54.0	248.9	49.5	244.0	10.3	236.8	1.8	260.2	0.8	264.9	2.1
84	241.4	123.6	243.8	94.3	251.1	53.3	249.8	49.3	244.6	10.5	237.1	1.8	260.6	0.5	265.3	2.2
85	243.5	124.0	245.5	94.6	252.1	54.0	250.1	49.5	244.4	10.3	237.1	1.6	260.6	0.5	265.4	1.9
86	245.9	124.3	247.7	94.9	252.8	53.7	250.7	49.5	244.1	10.3	237.0	1.9	260.1	0.3	265.2	1.9
87	247.7	124.4	249.2	94.9	253.6	53.7	250.8	49.6	244.3	10.3	237.0	1.9	260.1	0.5	265.7	1.8
88	250.1	124.8	251.2	95.0	254.2	54.1	251.6	49.9	244.6	10.7	237.4	1.9	260.6	0.1	265.7	2.1
89	251.9	124.4	252.6	94.7	254.9	54.0	251.7	49.9	244.6	10.7	237.1	2.1	260.6	0.5	265.6	1.6
90	254.3	124.5	254.6	94.7	255.3	54.1	252.6	49.9	245.0	10.7	237.4	2.1	260.5	0.5	265.7	1.8
91	256.3	124.5	256.7	95.1	255.3	54.1	252.6	49.9	244.9	10.8	237.3	2.3	260.6	0.1	265.8	1.6
92	258.4	124.1	257.8	94.6	256.4	53.7	253.2	50.0	245.3	10.8	237.3	2.1	260.8	0.3	265.8	1.6
93	261.0	124.1	260.1	94.4	256.9	54.0	254.0	50.2	245.3	11.1	237.2	2.6	260.8	0.3	265.7	1.5
94	262.8	123.7	261.9	94.4	257.9	54.1	254.8	50.7	245.2	11.2	237.6	2.7	260.6	0.3	265.7	1.4
95	265.2	123.3	263.5	94.4	259.1	54.5	255.7	50.7	245.4	12.0	237.8	3.3	260.7	0.3	266.1	1.5
96	267.1	123.2	265.5	94.3	259.1	54.5	255.8	50.9	245.4	12.6	237.5	3.7	260.5	0.4	266.1	1.5
97	269.2	123.0	267.4	94.3	259.6	54.5	257.0	50.6	245.4	12.0	237.5	4.1	260.5	0.4	265.7	1.6
98	271.2	122.2	269.0	94.0	260.5	54.5	257.4	50.6	245.7	12.6	237.7	4.8	260.7	0.3	265.7	1.6
99	273.6	122.3	271.2	94.3	262.2	54.5	258.8	50.8	245.7	12.6	237.8	5.3	260.3	0.3	265.7	1.5
100	275.5	121.8	273.3	93.6	265.0	54.7	261.9	51.0	246.1	13.3	237.8	5.3	260.3	0.4	265.7	1.5
101	277.8	121.8	275.3	93.6	265.0	54.5	261.0	51.1	246.7	13.3	238.2	6.1	260.7	0.3	265.7	1.6
102	280.2	121.4	278.1	93.6	267.2	54.5	263.5	51.3	247.1	14.1	238.6	6.8	260.5	0.3	265.9	1.4
103	282.7	121.4	280.3	93.5	269.0	54.1	265.5	51.3	247.8	14.8	238.6	7.8	260.5	0.3	265.7	1.4
104	284.8	120.6	282.7	92.9	271.0	54.0	268.0	51.0	249.2	15.0	239.3	8.6	260.4	0.0	265.9	1.5
105	287.3	120.6	285.0	92.7	274.0	53.3	270.3	51.0	249.7	15.0	241.5	10.1	260.8	0.1	265.8	1.4
106	289.9	120.6	287.7	92.7	277.1	53.3	273.4	51.0	249.7	16.8	241.5	11.8	260.7	0.5	265.9	1.4
107	292.7	120.1	290.8	92.4	280.4	52.6	276.9	50.3	251.1	18.0	242.2	13.4	260.9	0.3	266.1	1.6
108	295.1	119.9	293.2	92.0	284.0	52.6	279.9	50.3	254.4	18.7	243.8	15.4	260.9	0.3	266.1	1.4
109	297.8	119.9	296.0	91.6	288.1	52.2	284.1	49.5	256.6	19.7	245.1	17.6	260.5	0.4	265.9	1.2
110	300.4	119.9	299.0	90.8	292.2	52.2	288.4	49.1	259.4	21.1	248.1	19.9	260.8	0.9	266.5	1.2
111	302.5	119.9	301.6	91.0	296.7	51.7	292.6	49.1	262.1	22.0	250.8	22.7	261.0	0.9	266.6	0.8
112	304.7	119.7	304.2	90.8	300.6	51.4	297.1	49.2	265.6	23.8	254.3	25.1	262.2	3.3	266.7	0.7
113	306.6	119.9	306.7	90.6	304.9	51.3	301.4	48.8	269.3	25.0	258.1	26.6	264.3	4.4	268.2	0.8

Table A.4 Filtered Coordinate Data—Fourth-Order Low-Pass Filter, F_c = 5 Hz

FILTERED DATA LISTING:

FRAME	BASE RIB		R. HIP		R. KNEE		R. FIBUL		R. ANKLE		R. HEEL		R. META		R. TOE	
	X	Y	X	Y	X	Y	X	Y	X	Y	X	Y	X	Y	X	Y
-1	39.1	120.4	42.7	91.3	59.4	54.7	59.1	49.7	75.8	14.3	74.4	2.9	95.5	12.6	99.3	17.0
0	40.9	120.2	44.8	91.1	61.4	54.5	61.1	49.6	76.9	13.8	75.2	2.7	96.4	11.6	100.3	16.0
1	43.2	120.1	47.2	91.0	63.8	54.4	63.5	49.5	78.1	13.3	75.9	2.5	97.3	10.4	101.4	14.5
2	45.8	119.9	49.9	90.9	66.5	54.4	66.2	49.5	79.3	12.7	76.5	2.3	98.2	8.9	102.4	12.8
3	48.6	119.8	52.8	90.9	69.4	54.5	69.1	49.6	80.5	12.2	76.9	2.3	99.1	7.3	103.4	10.9
4	51.5	119.8	55.9	91.0	72.4	54.5	72.0	49.7	81.7	11.8	77.3	2.3	99.8	5.8	104.3	9.1
5	54.5	119.8	58.9	91.2	75.4	54.6	74.9	49.8	82.7	11.5	77.6	2.3	100.4	4.5	105.0	7.4
6	57.4	119.9	61.8	91.5	78.2	54.7	77.6	49.9	83.5	11.3	77.9	2.3	100.8	3.4	105.6	5.9
7	60.1	120.1	64.6	91.8	80.7	54.8	80.0	50.0	84.1	11.1	77.9	2.4	101.0	2.5	105.9	4.8
8	62.7	120.3	67.2	92.1	82.9	54.9	82.1	50.1	84.5	11.0	78.1	2.4	101.2	1.9	106.1	3.9
9	65.2	120.7	69.7	92.5	84.8	54.9	83.8	50.2	84.8	11.0	78.1	2.5	101.2	1.6	106.2	3.3
10	67.7	121.1	72.1	92.8	86.4	54.9	85.2	50.3	84.9	11.0	78.0	2.5	101.1	1.3	106.2	3.0
11	70.1	121.5	74.3	93.2	87.6	54.9	86.4	50.3	84.9	11.0	78.0	2.5	101.1	1.2	106.2	2.8
12	72.4	122.0	76.5	93.6	88.7	54.8	87.3	50.3	84.9	11.1	77.9	2.5	100.9	1.2	106.1	2.7
13	74.8	122.5	78.6	94.0	89.6	54.7	88.1	50.2	84.9	11.1	77.9	2.5	100.8	1.2	106.1	2.6
14	77.2	123.0	80.7	94.4	90.4	54.7	88.7	50.2	84.9	11.1	77.9	2.5	100.8	1.2	106.1	2.6
15	79.5	123.5	82.7	94.7	91.1	54.6	89.3	50.1	85.0	11.1	77.9	2.5	100.8	1.2	106.1	2.6
16	81.9	123.9	84.6	95.0	91.7	54.6	89.8	50.1	85.0	11.1	77.9	2.5	100.8	1.2	106.2	2.5
17	84.2	124.3	86.5	95.2	92.3	54.5	90.3	50.1	85.1	11.2	78.0	2.5	100.9	1.2	106.2	2.5
18	86.5	124.5	88.3	95.3	92.9	54.5	90.7	50.1	85.2	11.2	78.0	2.5	100.9	1.2	106.3	2.5
19	88.8	124.7	90.1	95.4	93.5	54.4	91.2	50.1	85.3	11.3	78.1	2.6	101.0	1.2	106.4	2.4
20	91.0	124.7	91.9	95.4	94.1	54.4	91.8	50.0	85.5	11.3	78.2	2.6	101.0	1.1	106.5	2.4
21	93.3	124.7	93.6	95.4	94.7	54.3	92.3	50.0	85.6	11.4	78.3	2.7	101.1	1.0	106.5	2.4
22	95.4	124.6	95.4	95.3	95.4	54.3	92.9	50.0	85.7	11.4	78.4	2.8	101.1	1.0	106.6	2.4
23	97.6	124.6	97.2	95.2	96.1	54.3	93.4	50.0	85.9	11.5	78.4	2.9	101.1	1.0	106.7	2.4
24	99.7	124.2	99.0	95.1	96.8	54.3	94.0	50.1	86.0	11.6	78.5	3.0	101.1	1.0	106.7	2.4
25	101.9	124.0	100.9	94.9	97.6	54.3	94.7	50.2	86.1	11.7	78.6	3.2	101.1	1.0	106.8	2.4
26	104.1	123.7	102.8	94.8	98.4	54.3	95.4	50.4	86.2	11.9	78.7	3.4	101.1	1.0	106.8	2.3
27	106.3	123.5	104.8	94.7	99.3	54.3	96.1	50.4	86.3	12.1	78.7	3.7	101.1	0.9	106.8	2.3
28	108.5	123.2	106.8	94.5	100.3	54.4	97.0	50.5	86.4	12.4	78.7	4.0	101.1	0.9	106.8	2.2
29	110.8	122.9	108.9	94.3	101.4	54.4	98.0	50.6	86.6	12.7	78.8	4.5	101.0	0.9	106.8	2.2
30	113.1	122.5	111.0	94.2	102.7	54.4	99.2	50.7	86.8	13.1	78.9	5.0	101.0	0.9	106.8	2.2
31	115.5	122.2	113.2	94.0	104.1	54.4	100.6	50.7	87.1	13.6	79.0	5.6	101.0	0.9	106.8	2.2
32	117.8	121.9	115.5	93.8	105.8	54.4	102.1	50.9	87.6	14.1	79.2	6.3	101.0	0.9	106.8	2.2
33	120.2	121.5	117.8	93.6	107.6	54.4	103.9	51.0	88.1	14.6	79.4	7.1	101.0	1.0	106.8	2.2

Table A.4 (Continued)

FILTERED DATA LISTING:

FRAME	BASE RIB X	Y	R. HIP X	Y	R. KNEE X	Y	R. FIBUL X	Y	R. ANKLE X	Y	R. HEEL X	Y	R. META X	Y	R. TOE X	Y
34	122.7	121.2	120.2	93.1	109.8	54.3	106.0	51.0	88.8	15.3	79.8	8.2	101.0	0.9	106.8	2.2
35	125.3	120.9	122.8	93.1	112.2	54.2	108.3	50.9	89.7	16.0	80.3	9.4	100.9	0.9	106.8	2.2
36	127.9	120.7	125.4	92.9	114.9	53.9	111.0	50.8	90.8	16.8	81.1	10.9	100.8	1.0	106.7	2.1
37	130.5	120.4	128.1	92.6	118.0	53.6	114.0	50.6	92.1	17.6	82.1	12.6	100.8	1.0	106.6	2.1
38	133.2	120.2	130.9	92.2	121.3	53.2	117.4	50.3	93.7	18.6	83.4	14.6	100.7	1.1	106.5	2.0
39	135.8	120.1	133.7	92.1	125.0	52.8	121.0	50.3	95.6	19.7	85.1	16.8	100.8	1.8	106.4	1.9
40	138.4	120.0	136.6	91.7	129.0	52.4	125.0	49.7	97.9	20.9	87.2	19.1	101.1	1.8	106.6	1.8
41	141.0	120.0	139.4	91.4	133.2	52.0	129.3	49.4	100.6	22.1	89.7	21.3	101.8	2.4	107.2	1.8
42	143.4	120.0	142.3	91.3	137.6	51.7	133.7	49.4	103.7	23.4	92.6	23.5	103.2	3.2	108.3	1.8
43	145.8	120.2	145.1	91.2	142.2	51.5	138.3	49.1	107.1	24.6	96.0	25.3	105.4	4.0	110.2	1.9
44	148.0	120.4	147.8	91.2	146.8	51.5	143.0	49.0	110.9	25.6	99.7	26.6	108.5	4.8	113.1	2.1
45	150.2	120.6	150.4	91.4	151.3	51.6	147.6	49.0	115.1	26.3	103.8	27.5	112.5	5.4	117.0	2.3
46	152.4	121.0	153.0	91.6	155.8	51.9	152.2	49.4	119.5	26.8	108.1	27.5	117.5	5.8	121.8	2.6
47	154.5	121.4	155.5	92.0	160.2	52.4	156.6	49.8	124.2	26.8	112.8	27.0	123.2	5.8	127.5	2.7
48	156.6	121.8	158.0	92.4	164.4	53.1	160.9	50.4	129.1	26.5	117.7	25.9	129.5	5.8	134.0	2.8
49	158.7	122.3	160.5	92.9	168.4	53.8	165.1	50.9	134.2	25.8	122.5	24.3	136.3	5.5	141.0	2.7
50	160.8	122.9	162.9	93.4	172.3	54.5	169.0	51.6	139.6	24.7	128.5	22.2	143.5	5.0	148.4	2.6
51	162.9	123.3	165.4	93.9	175.9	55.2	172.9	52.2	145.2	23.4	134.5	19.9	151.0	4.3	156.1	2.4
52	165.0	123.7	167.8	94.3	179.4	55.9	176.5	52.7	151.0	22.0	140.5	17.3	158.7	3.7	164.0	2.2
53	167.2	124.1	170.1	94.7	182.8	56.5	180.0	53.2	157.0	20.4	146.9	14.7	166.5	3.0	171.9	2.1
54	169.4	124.3	172.4	94.9	185.9	57.0	183.3	53.6	163.2	18.8	153.6	12.1	174.3	2.5	179.8	2.0
55	171.5	124.5	174.7	95.1	188.9	57.4	186.4	53.9	169.6	17.3	160.5	9.8	182.0	2.1	187.7	2.2
56	173.7	124.6	177.0	95.2	191.7	57.6	189.4	54.0	176.0	16.0	167.5	7.7	189.7	1.9	195.5	2.9
57	176.0	124.6	179.2	95.2	194.4	57.7	192.2	53.9	182.5	14.9	174.7	5.9	197.3	1.9	203.1	2.9
58	178.2	124.5	181.3	95.0	196.9	57.6	194.9	53.8	188.9	14.0	181.8	4.5	204.7	2.2	210.4	3.6
59	180.4	124.4	183.5	94.8	199.3	57.3	197.5	53.5	195.3	13.0	188.9	3.5	211.9	2.7	217.5	4.5
60	182.7	124.2	185.6	94.6	201.5	57.0	200.0	53.5	201.4	13.0	195.8	2.9	218.8	3.5	224.3	5.7
61	184.9	124.0	187.8	94.2	203.7	56.7	202.4	52.7	207.3	12.8	202.4	2.5	225.3	4.5	230.6	7.1
62	187.2	123.6	190.0	93.8	205.5	56.3	204.8	52.2	212.7	13.0	208.6	2.3	231.3	5.7	236.4	8.6
63	189.5	123.3	192.2	93.5	208.1	56.1	207.2	51.7	217.8	13.0	214.3	2.2	236.9	7.1	241.6	10.3
64	191.7	122.9	194.5	93.5	210.4	55.9	209.7	51.2	222.3	13.4	219.5	2.2	241.8	8.3	246.3	11.9
65	194.1	122.5	197.0	92.7	212.8	55.5	212.2	50.8	226.3	13.4	223.9	2.2	246.0	9.4	250.3	13.2
66	196.4	122.1	199.5	92.4	215.2	55.2	214.8	50.5	229.7	13.4	227.6	2.2	249.5	10.1	253.6	14.1
67	198.8	121.8	202.1	92.1	217.6	55.0	217.5	50.2	232.6	13.1	230.6	2.1	252.4	10.1	256.5	14.5
68	201.3	121.5	204.8	92.1	220.6	54.6	220.2	49.9	234.9	12.7	232.8	2.0	254.6	9.4	258.5	14.2
69	203.8	121.2	207.5	91.9	223.3	54.4	222.9	49.8	236.8	12.2	234.4	1.8	256.2	9.4	260.2	13.4
70	206.4	121.1	210.3	91.7	226.1	54.3	225.3	49.6	238.3	12.0	235.5	1.6	257.4	8.3	261.5	12.2
71	209.0	120.8	213.1	91.6	228.8	54.3	228.3	49.5	239.5	11.8	235.9	1.5	258.2	7.0	262.4	10.6
72	211.7	120.7	215.9	91.6	231.5	54.3	231.0	49.5	240.5	11.3	236.2	1.4	258.8	5.7	263.1	9.0

Table A.4 (Continued)

FILTERED DATA LISTING:

FRAME	BASE RIB X	Y	R. HIP X	Y	R. KNEE X	Y	R. FIBUL X	Y	R. ANKLE X	Y	R. HEEL X	Y	R. META X	Y	R. TOE X	Y
73	214.3	120.6	218.6	91.7	234.2	54.3	233.5	49.5	241.3	11.0	236.4	1.4	259.2	4.4	263.5	7.3
74	217.0	120.6	221.3	91.9	236.7	54.3	236.0	49.6	241.9	10.7	236.4	1.4	259.5	3.2	263.9	5.8
75	219.6	120.7	224.0	92.1	239.1	54.4	238.3	49.7	242.5	10.5	236.4	1.5	259.6	2.3	264.1	4.6
76	222.1	120.8	226.5	92.3	241.2	54.4	240.4	49.7	242.9	10.4	236.5	1.5	259.8	1.6	264.3	3.6
77	224.6	121.0	229.0	92.6	243.2	54.4	242.2	49.7	243.2	10.3	236.5	1.6	259.9	1.1	264.4	2.9
78	227.1	121.3	231.4	92.8	244.9	54.3	243.8	49.6	243.5	10.2	236.6	1.6	259.9	0.9	264.6	2.5
79	229.6	121.7	233.7	93.1	246.4	54.3	245.2	49.6	243.7	10.2	236.6	1.7	260.0	0.7	264.7	2.2
80	232.0	122.0	235.9	93.3	247.7	54.2	246.4	49.6	243.8	10.2	236.7	1.7	260.1	0.7	264.8	2.1
81	234.4	122.2	238.0	93.6	248.8	54.2	247.4	49.5	243.9	10.2	236.7	1.7	260.2	0.7	264.9	2.0
82	236.7	122.8	240.0	93.8	249.8	54.1	248.2	49.5	244.0	10.2	236.8	1.8	260.3	0.7	265.0	2.0
83	239.1	123.2	242.0	94.0	250.7	54.0	249.2	49.5	244.2	10.3	236.9	1.8	260.3	0.6	265.1	2.0
84	241.3	123.6	243.9	94.3	251.4	54.0	249.6	49.5	244.2	10.3	237.0	1.8	260.4	0.6	265.2	2.0
85	243.6	123.9	245.7	94.5	252.1	53.9	250.2	49.5	244.4	10.4	237.0	1.8	260.4	0.5	265.3	2.0
86	245.7	124.2	247.6	94.7	252.7	53.9	250.8	49.5	244.4	10.4	237.1	1.8	260.5	0.5	265.4	1.9
87	247.9	124.4	249.3	94.8	253.2	53.9	251.1	49.6	244.5	10.5	237.2	1.8	260.5	0.4	265.5	1.9
88	250.0	124.5	251.1	94.9	253.7	53.9	251.5	49.6	244.6	10.5	237.2	1.8	260.5	0.4	265.5	1.8
89	252.1	124.5	252.8	94.9	254.1	53.9	251.8	49.7	244.7	10.6	237.3	1.9	260.5	0.4	265.6	1.8
90	254.3	124.5	254.5	94.9	254.7	54.0	252.2	49.8	244.8	10.7	237.3	1.9	260.5	0.3	265.7	1.7
91	256.4	124.4	256.3	94.8	255.2	54.0	252.6	49.9	244.9	10.8	237.3	2.0	260.6	0.3	265.7	1.7
92	258.5	124.2	258.1	94.8	255.7	54.0	253.0	50.0	245.0	10.9	237.3	2.1	260.6	0.3	265.7	1.6
93	260.7	124.0	259.9	94.7	256.3	54.1	253.5	50.1	245.1	11.1	237.4	2.3	260.6	0.3	265.8	1.6
94	262.8	123.7	261.7	94.6	257.0	54.1	254.1	50.2	245.2	11.2	237.4	2.5	260.6	0.3	265.8	1.5
95	264.9	123.4	263.5	94.5	257.8	54.2	254.7	50.4	245.3	11.4	237.4	2.8	260.6	0.3	265.8	1.5
96	267.1	123.1	265.4	94.4	258.7	54.3	255.5	50.5	245.4	11.7	237.5	3.2	260.6	0.3	265.8	1.5
97	269.2	122.8	267.3	94.3	259.6	54.4	256.4	50.7	245.5	12.0	237.5	3.6	260.6	0.3	265.8	1.5
98	271.3	122.5	269.2	94.2	260.8	54.4	257.4	50.8	245.7	12.3	237.6	4.1	260.6	0.3	265.8	1.5
99	273.5	122.2	271.3	94.0	262.1	54.5	258.7	50.9	245.9	12.7	237.7	4.7	260.5	0.2	265.8	1.5
100	275.7	121.9	273.4	93.9	263.6	54.5	260.2	51.1	246.3	13.1	237.9	5.3	260.5	0.2	265.8	1.5
101	278.0	121.7	275.7	93.7	265.4	54.4	261.9	51.1	246.8	13.6	238.2	6.1	260.5	0.2	265.8	1.5
102	280.4	121.4	278.1	93.5	267.4	54.3	263.9	51.2	247.4	14.2	238.7	7.0	260.6	0.2	265.8	1.5
103	282.9	121.1	280.6	93.3	269.7	54.2	266.1	51.1	248.2	14.8	239.2	8.1	260.6	0.2	265.9	1.5
104	285.3	120.9	283.1	93.0	272.2	53.9	268.5	51.0	249.2	15.4	239.9	9.3	260.6	0.2	265.9	1.4
105	287.8	120.6	285.6	92.8	274.9	53.7	271.2	50.9	250.3	16.2	240.7	10.6	260.6	0.2	266.0	1.4
106	290.1	120.4	288.0	92.5	277.5	53.4	273.8	50.6	251.4	16.9	241.5	12.0	260.6	0.3	266.0	1.4
107	292.2	120.2	290.2	92.3	280.1	53.2	276.3	50.6	252.6	17.6	242.3	13.3	260.6	0.3	266.1	1.4
108	293.9	120.1	292.1	92.1	282.4	52.9	278.5	50.2	253.7	18.2	243.1	14.5	260.5	0.4	266.1	1.4
109	295.3	120.0	293.6	92.1	284.2	52.7	280.3	50.2	254.6	18.7	243.8	15.5	260.5	0.4	266.1	1.4
110	296.3	119.9	294.6	91.7	285.7	52.6	281.8	49.9	255.4	19.1	244.5	16.3	260.5	0.5	266.2	1.4
111	297.1	119.9	295.6	91.6	287.0	52.5	283.1	49.8	256.2	19.5	245.2	17.1	260.6	0.6	266.2	1.3

158

Table A.5 Displacement/Velocity/Acceleration—C of G of Foot

POS	TIME S	THETA DEG	OMEGA R/S	ALPHA R/S/S	CGX M	VX M/S	AX M/S/S	CGY M	VY M/S	AY M/S/S	MODE1	MODE2
1	0.000	351.403	-2.556	-41.459	0.877	0.636	0.167	0.118	-0.574	-3.457	HSR	1
2	0.017	348.615	-3.190	-30.430	0.888	0.628	-1.435	0.108	-0.609	-0.599		1
3	0.033	345.309	-3.571	-12.501	0.898	0.588	-3.355	0.098	-0.594	2.150		1
4	0.050	341.794	-3.607	7.693	0.907	0.516	-5.033	0.088	-0.537	4.335		1
5	0.067	338.419	-3.314	24.364	0.915	0.420	-6.050	0.080	-0.450	5.630		1
6	0.083	335.463	-2.795	34.193	0.921	0.314	-6.255	0.073	-0.349	5.966		1
7	0.100	333.080	-2.174	36.862	0.926	0.212	-5.749	0.068	-0.251	5.544		1
8	0.117	331.309	-1.566	33.664	0.928	0.123	-4.745	0.065	-0.165	4.652		1
9	0.133	330.089	-1.052	26.853	0.930	0.053	-3.482	0.063	-0.096	3.540		1
10	0.150	329.300	-0.670	19.031	0.930	0.007	-2.169	0.062	-0.047	2.422	TOL	1
11	0.167	328.809	-0.417	12.049	0.930	-0.019	-0.960	0.061	-0.015	1.461		1
12	0.183	328.502	-0.269	6.610	0.929	-0.025	0.018	0.061	-0.002	0.719		1
13	0.200	328.295	-0.197	2.701	0.929	-0.018	0.667	0.061	0.009	0.210		1
14	0.217	328.126	-0.179	0.135	0.929	-0.003	0.963	0.061	0.009	-0.060		1
15	0.233	327.954	-0.193	-1.246	0.929	0.014	0.970	0.062	0.007	-0.139		1
16	0.250	327.758	-0.220	-1.769	0.929	0.029	0.792	0.062	0.005	-0.127		1
17	0.267	327.533	-0.252	-1.982	0.930	0.040	0.559	0.062	0.003	-0.086		1
18	0.283	327.278	-0.286	-2.180	0.931	0.048	0.373	0.062	0.002	-0.011		1
19	0.300	326.987	-0.324	-2.229	0.931	0.053	0.218	0.062	0.002	-0.110		1
20	0.317	326.659	-0.361	-2.165	0.932	0.055	0.012	0.062	0.005	0.245		1
21	0.333	326.298	-0.396	-2.360	0.933	0.053	-0.234	0.062	0.010	0.345		1
22	0.350	325.901	-0.439	-2.982	0.934	0.047	-0.398	0.062	0.017	0.400		1
23	0.367	325.459	-0.496	-1.917	0.935	0.040	-0.412	0.063	0.024	0.440		1
24	0.383	324.954	-0.570	-5.137	0.936	0.033	-0.345	0.063	0.031	0.504		1
25	0.400	324.370	-0.667	-6.679	0.936	0.028	-0.263	0.064	0.041	0.621		1
26	0.417	323.680	-0.793	-8.479	0.936	0.025	-0.090	0.064	0.052	0.788		1
27	0.433	322.856	-0.950	-10.589	0.937	0.034	0.288	0.065	0.057	0.963		1
28	0.450	321.866	-1.146	-13.019	0.938	0.034	0.804	0.067	0.084	1.093		1
29	0.467	320.668	-1.384	-15.314	0.938	0.052	1.271	0.068	0.103	1.159		1
30	0.483	319.222	-1.656	-17.162	0.939	0.057	1.633	0.070	0.123	1.179		1
31	0.500	317.504	-1.956	-19.351	0.941	0.107	1.980	0.072	0.143	1.174		1

Table A.5 *(Continued)*

DISPLACEMENT/VELOCITY/ACCELERATION DATA GENERATED FROM THE DATA OF THE R. FOOT USING FINITE DIFFERENCE AFTER DIGITAL FILTERING

POS	TIME S	THETA DEG	OMEGA R/S	ALPHA R/S/S	CGX M	VX M/S	AX M/S/S	CGY M	VY M/S	AY M/S/S	MODE1	MODE2
32	0.517	315.486	-2.301	-23.410	0.943	0.143	2.367	0.075	0.162	1.159		1
33	0.533	313.108	-2.736	-30.133	0.945	0.186	2.714	0.078	0.181	1.187		1
34	0.550	310.259	-3.306	-38.403	0.949	0.233	2.958	0.081	0.202	1.359		1
35	0.567	306.793	-4.017	-45.349	0.953	0.284	3.253	0.084	0.227	1.778		1
36	0.583	302.585	-4.818	-47.505	0.958	0.342	3.976	0.088	0.261	2.460	HSL	1
37	0.600	297.589	-5.601	-42.039	0.964	0.417	5.524	0.093	0.309	3.308		1
38	0.617	291.887	-6.219	-27.424	0.972	0.526	8.045	0.099	0.371	4.110		1
39	0.633	285.709	-6.515	-3.717	0.982	0.685	11.341	0.105	0.446	4.503		1
40	0.650	279.441	-6.343	26.637	0.995	0.904	14.971	0.114	0.521	4.021		1
41	0.667	273.592	-5.627	58.193	1.012	1.184	18.367	0.123	0.580	2.341		1
42	0.683	268.693	-4.403	84.207	1.034	1.516	20.962	0.133	0.599	-0.439		1
43	0.700	265.180	-2.819	99.874	1.063	1.883	22.326	0.143	0.565	-3.778		1
44	0.717	263.307	-1.073	103.916	1.097	2.261	22.267	0.152	0.473	-6.985		1
45	0.733	263.130	0.645	97.937	1.138	2.625	20.868	0.159	0.332	-9.507	TOF	1
46	0.750	264.539	2.192	85.337	1.185	2.956	18.520	0.164	0.156	-11.021		1
47	0.767	267.317	3.490	69.761	1.237	3.243	15.800	0.163	-0.035	-11.406		1
48	0.783	271.207	4.518	53.395	1.293	3.483	13.214	0.162	-0.224	-10.702		1
49	0.800	275.947	5.270	37.070	1.353	3.683	10.991	0.156	-0.392	-9.040		1
50	0.817	281.274	5.754	21.634	1.416	3.850	9.066	0.149	-0.525	-6.619		1
51	0.833	286.937	5.992	8.327	1.481	3.986	7.258	0.139	-0.613	-3.758		1
52	0.850	292.720	6.031	-2.147	1.549	4.092	5.462	0.128	-0.651	-0.867		1
53	0.867	298.458	5.920	-10.020	1.618	4.168	3.712	0.117	-0.642	1.728		1
54	0.884	304.029	5.697	-15.725	1.688	4.215	2.043	0.107	-0.593	3.928		1
55	0.900	309.341	5.396	-19.318	1.758	4.236	0.363	0.097	-0.511	5.783		1
56	0.917	314.336	5.053	-20.583	1.829	4.227	-0.363	0.090	-0.400	7.125		1
57	0.934	318.994	4.770	-19.325	1.899	4.186	-1.500	0.084	-0.266	8.509		1
58	0.950	323.333	4.409	-15.629	1.968	4.104	-3.693	0.081	-0.116	9.236		1
59	0.967	327.415	4.189	-10.362	2.036	3.978	-6.231	0.082	0.041	9.335		1
60	0.984	331.334	4.063	-5.750	2.101	3.804	-8.996	0.082	0.195	8.542		1
61	1.000	335.177	3.997	-4.977	2.163	3.583	-11.852	0.087	0.326	6.604		1
62	1.017	338.969	3.897	-10.404	2.240	3.314	-14.709	0.093	0.415	3.513		1
63	1.034	342.622	3.650	-21.815	2.273	3.001	-17.461	0.100	0.443	-0.393		1
64	1.050	345.942	3.170	-36.416	2.320	2.650	-19.907	0.108	0.402	-4.515		1
65	1.067	348.678	2.436	-50.302	2.362	2.276	-21.729	0.114	0.293	-8.120		1
66	1.084	350.595	1.493	-59.965	2.396	1.898	-22.577	0.118	0.131	-10.468		1
67	1.100	351.429	0.437	-63.535	2.425	1.535	-20.674	0.118	-0.056	-11.062	HSR	1
68	1.117	351.530	-0.619	-60.347	2.447	1.208	-18.165	0.116	-0.238	-9.897		1
69	1.134	350.347	-1.575	-52.199	2.465	0.930	-15.031	0.110	-0.386	-7.418		1
70	1.150	348.420	-2.360	-40.118	2.478	0.707	-11.740	0.103	-0.485	-4.252		1

Table A.5 (Continued)

POS	TIME s	THETA DEG	OMEGA R/S	ALPHA R/S/S	CGX M	VX M/S	AX M/S/S	CGY M	VY M/S	AY M/S/S	MODE1	MODE2
71	1.167	345.840	-2.913	-24.515	2.489	0.538	-8.837	0.094	-0.528	-0.994		1
72	1.184	342.855	-3.177	-6.333	2.496	0.413	-6.643	0.085	-0.518	1.845		1
73	1.200	339.771	-3.144	11.501	2.502	0.317	-5.061	0.077	-0.466	3.911		1
74	1.217	336.888	-2.773	25.375	2.507	0.244	-3.854	0.070	-0.388	5.082		1
75	1.234	334.435	-2.278	33.274	2.511	0.188	-2.938	0.064	-0.297	5.412		1
76	1.250	332.537	-1.684	35.018	2.513	0.146	-2.273	0.060	-0.207	5.024		1
77	1.267	331.218	-1.110	31.355	2.515	0.113	-1.727	0.057	-0.129	4.117		1
78	1.284	330.415	-0.639	23.787	2.517	0.088	-1.197	0.055	-0.070	2.972		1
79	1.300	329.998	-0.317	14.500	2.518	0.073	-0.727	0.055	-0.030	1.852		1
80	1.317	329.809	-0.155	5.567	2.519	0.064	-0.403	0.054	-0.008	0.949	TOI	1
81	1.334	329.701	-0.132	-1.477	2.520	0.059	-0.240	0.054	0.001	0.370		1
82	1.350	329.557	-0.204	-5.291	2.521	0.056	-0.206	0.054	0.004	0.079		1
83	1.367	329.311	-0.308	-5.301	2.522	0.052	-0.295	0.055	0.004	-0.058		1
84	1.384	328.969	-0.381	-2.876	2.523	0.046	-0.391	0.055	0.002	-0.117		1
85	1.400	328.583	-0.404	-0.503	2.524	0.039	-0.348	0.055	-0.000	-0.080		1
86	1.417	328.197	-0.398	0.405	2.525	0.035	-0.158	0.055	-0.001	-0.062		1
87	1.434	327.822	-0.391	0.144	2.525	0.034	0.027	0.055	0.002	0.250		1
88	1.450	327.450	-0.393	-0.445	2.526	0.036	0.067	0.055	0.008	0.392		1
89	1.467	327.071	-0.405	-1.076	2.526	0.036	-0.018	0.055	0.015	0.452		1
90	1.484	326.676	-0.429	-1.844	2.527	0.035	-0.018	0.055	0.023	0.461		1
91	1.500	326.251	-0.467	-2.634	2.527	0.032	-0.173	0.056	0.031	0.466		1
92	1.517	325.784	-0.517	-3.240	2.528	0.029	-0.188	0.057	0.038	0.482		1
93	1.534	325.264	-0.575	-3.950	2.528	0.026	-0.148	0.057	0.047	0.511		1
94	1.550	324.685	-0.649	-5.358	2.529	0.024	-0.018	0.058	0.055	0.555		1
95	1.567	324.025	-0.754	-7.668	2.529	0.026	0.208	0.059	0.065	0.603		1
96	1.584	323.246	-0.904	-10.698	2.530	0.031	0.532	0.060	0.076	0.654		1
97	1.600	322.298	-1.110	-14.077	2.530	0.043	0.945	0.061	0.087	0.743		1
98	1.617	321.125	-1.374	-17.225	2.531	0.063	1.413	0.063	0.100	0.876		1
99	1.634	319.674	-1.685	-19.883	2.532	0.090	1.907	0.065	0.116	1.001		1
100	1.650	317.907	-2.036	-22.571	2.534	0.126	2.389	0.067	0.134	1.087		1
101	1.667	315.784	-2.437	-25.596	2.537	0.170	2.782	0.069	0.152	1.180		1
102	1.684	313.251	-2.890	-28.079	2.540	0.219	2.965	0.072	0.173	1.299		1
103	1.700	310.203	-3.373	-28.454	2.544	0.269	2.794	0.075	0.196	1.324		1
104	1.717	306.808	-3.838	-24.860	2.549	0.312	2.163	0.078	0.217	1.033		1
105	1.734	302.931	-4.202	-15.037	2.554	0.341	1.072	0.082	0.230	0.323	HSL	1
106	1.750	298.781	-4.340	-1.901	2.560	0.348	-0.316	0.086	0.228	-0.624		1
107	1.767	294.641	-4.139	22.258	2.566	0.331	-1.639	0.090	0.209	-1.432		1
108	1.784	290.875	-3.548	37.182	2.571	0.293	-2.311	0.093	0.180	-1.794		1
109	1.800	287.768	-2.899	34.361	2.576	0.253	-1.373	0.096	0.152	-0.972		1

Table A.6 Displacement/Velocity/Acceleration—C of G of Shank

DISPLACEMENT/VELOCITY/ACCELERATION DATA GENERATED FROM THE DATA OF THE R. SHANK USING FINITE DIFFERENCE AFTER DIGITAL FILTERING

POS	TIME S	THETA DEG	OMEGA R/S	ALPHA R/S/S	CGX M	VX M/S	AX M/S/S	CGY M	VY M/S	AY M/S/S	MODE1	MODE2
1	0.000	289.084	-1.983	-21.380	0.700	1.180	6.889	0.366	-0.174	1.022	HSP	1
2	0.017	287.019	-2.299	-16.530	0.721	1.275	4.015	0.363	-0.146	2.067		1
3	0.033	284.693	-2.534	-11.682	0.742	1.313	0.510	0.361	-0.105	2.565		1
4	0.050	282.179	-2.688	-6.466	0.764	1.292	-2.988	0.360	-0.061	2.421		1
5	0.067	279.558	-2.749	-0.479	0.785	1.214	-6.002	0.359	-0.024	1.758		1
6	0.083	276.927	-2.704	5.702	0.805	1.092	-8.166	0.359	-0.002	0.867		1
7	0.100	274.392	-2.559	10.901	0.822	0.942	-9.263	0.359	-0.005	0.102		1
8	0.117	272.038	-2.341	14.291	0.836	0.783	-9.267	0.359	0.001	-0.313		1
9	0.133	269.921	-2.083	15.719	0.848	0.633	-8.357	0.359	-0.006	-0.391		1
10	0.150	268.060	-1.817	15.528	0.857	0.504	-6.862	0.359	-0.012	-0.281	TOL	1
11	0.167	266.450	-1.565	14.318	0.865	0.404	-5.152	0.359	-0.015	-0.111		1
12	0.183	265.070	-1.339	12.646	0.871	0.332	-3.556	0.359	-0.016	-0.048		1
13	0.200	263.892	-1.143	10.802	0.876	0.285	-2.286	0.358	-0.014	0.153		1
14	0.217	262.886	-0.979	8.850	0.880	0.256	-1.395	0.358	-0.010	0.178		1
15	0.233	262.021	-0.848	6.787	0.884	0.239	-0.806	0.358	-0.008	0.109		1
16	0.250	261.265	-0.753	4.637	0.888	0.229	-0.397	0.358	-0.007	-0.020		1
17	0.267	260.583	-0.694	2.545	0.892	0.225	-0.042	0.358	-0.008	-0.110		1
18	0.283	259.940	-0.668	0.745	0.896	0.228	0.311	0.358	-0.011	-0.061		1
19	0.300	259.307	-0.669	-0.659	0.900	0.236	0.593	0.357	-0.006	0.146		1
20	0.317	258.662	-0.690	-1.704	0.904	0.248	0.687	0.357	-0.010	0.410		1
21	0.333	257.989	-0.726	-2.405	0.908	0.259	0.594	0.357	-0.006	0.587		1
22	0.350	257.275	-0.770	-2.828	0.912	0.267	0.474	0.357	-0.003	0.620		1
23	0.367	256.517	-0.820	-3.223	0.917	0.275	0.502	0.358	0.014	0.574		1
24	0.383	255.709	-0.878	-3.887	0.921	0.284	0.716	0.358	0.024	0.562		1
25	0.400	254.841	-0.950	-4.906	0.926	0.298	1.067	0.358	0.033	0.647		1
26	0.417	253.895	-1.041	-6.040	0.931	0.320	1.570	0.360	0.043	0.803		1
27	0.433	252.852	-1.151	-6.968	0.937	0.351	2.278	0.361	0.055	0.952		1
28	0.450	251.696	-1.274	-7.643	0.943	0.396	3.132	0.362	0.069	1.027		1
29	0.467	250.419	-1.406	-8.281	0.950	0.455	4.003	0.363	0.086	0.988		1
30	0.483	249.011	-1.550	-9.083	0.958	0.529	4.854	0.365	0.104	0.808		1
31	0.500	247.459	-1.709	-10.044	0.968	0.617	5.754	0.367	0.119	0.487		1
32	0.517	245.747	-1.885	-11.051	0.979	0.721	6.751	0.370	0.131	-0.110		1
33	0.533	243.859	-2.077	-12.043	0.992	0.842	7.769	0.372	0.134	-0.190		1

162

Table A.6 (Continued)

DISPLACEMENT/VELOCITY/ACCELERATION DATA GENERATED FROM THE DATA OF THE B. SHANK USING FINITE DIFFERENCE AFTER DIGITAL FILTERING

POS	TIME S	THETA DEG	OMEGA R/S	ALPHA R/S/S	CGX M	VX M/S	AX M/S/S	CGY M	VY M/S	AY M/S/S	MODE1	MODE2
34	0.550	241.779	-2.286	-12.968	1.007	0.980	8.683	0.374	0.129	-0.325		1
35	0.567	239.492	-2.509	-13.479	1.024	1.132	9.449	0.376	0.123	-0.266		1
36	0.583	236.985	-2.735	-12.833	1.045	1.295	10.130	0.378	0.120	-0.016	HSL	1
37	0.600	234.267	-2.937	-10.404	1.068	1.469	10.785	0.380	0.123	0.385		1
38	0.617	231.375	-3.082	-6.228	1.094	1.655	11.359	0.382	0.133	0.863		1
39	0.633	228.379	-3.145	-0.622	1.123	1.848	11.633	0.385	0.152	1.319		1
40	0.650	225.367	-3.103	6.521	1.155	2.042	11.281	0.387	0.177	1.613		1
41	0.667	222.451	-2.927	15.477	1.191	2.224	10.113	0.391	0.205	1.603		1
42	0.683	219.775	-2.587	25.800	1.229	2.380	8.278	0.394	0.231	1.242		1
43	0.700	217.509	-2.067	36.068	1.270	2.500	6.223	0.398	0.247	0.585		1
44	0.717	215.826	-1.385	44.597	1.313	2.587	4.372	0.402	0.250	-0.295	TOR	1
45	0.733	214.864	-0.580	50.301	1.356	2.646	2.868	0.407	0.237	-1.342		1
46	0.750	214.717	0.292	52.834	1.401	2.683	1.658	0.410	0.205	-2.446		1
47	0.767	215.423	1.181	52.279	1.446	2.701	0.729	0.413	0.155	-3.432		1
48	0.783	216.973	2.035	48.986	1.491	2.707	0.163	0.416	0.091	-4.127		1
49	0.800	219.311	2.814	43.585	1.536	2.707	-0.033	0.416	0.018	-4.405		1
50	0.817	222.349	3.489	36.895	1.581	2.706	-0.033	0.416	-0.056	-4.214		1
51	0.833	225.975	4.044	29.782	1.626	2.706	-0.106	0.415	-0.123	-3.636		1
52	0.850	230.074	4.482	23.112	1.677	2.702	-0.420	0.412	-0.177	-2.872		1
53	0.867	234.536	4.815	17.538	1.716	2.692	-0.921	0.409	-0.218	-2.099		1
54	0.884	239.272	5.066	13.121	1.761	2.672	-1.477	0.405	-0.247	-1.332		1
55	0.900	244.214	5.252	9.341	1.805	2.642	-2.045	0.400	-0.263	-0.490		1
56	0.917	249.305	5.378	5.518	1.849	2.603	-2.680	0.396	-0.263	0.402		1
57	0.934	254.486	5.436	1.200	1.892	2.553	-3.431	0.392	-0.249	1.182		1
58	0.950	259.689	5.418	-3.836	1.934	2.489	-4.271	0.388	-0.224	1.710		1
59	0.967	264.835	5.308	-9.881	1.975	2.411	-5.094	0.384	-0.192	1.961		1
60	0.984	269.830	5.088	-17.321	2.015	2.319	-5.767	0.381	-0.159	1.952		1
61	1.000	274.555	4.731	-26.186	2.053	2.218	-6.202	0.379	-0.127	1.695		1
62	1.017	278.867	4.215	-35.761	2.089	2.112	-6.407	0.377	-0.102	1.222		1
63	1.034	282.607	3.539	-44.757	2.123	2.005	-6.465	0.376	-0.087	0.608		1
64	1.050	285.627	2.723	-51.806	2.156	1.897	-6.454	0.374	-0.082	-0.057		1
65	1.067	287.809	1.811	-55.756	2.186	1.790	-6.416	0.373	-0.088	-0.681		1
66	1.084	289.087	0.864	-55.750	2.215	1.683	-6.351	0.371	-0.105	-1.147		1
67	1.100	289.459	-0.047	-51.495	2.242	1.578	-6.230	0.369	-0.127	-1.500		1
68	1.117	288.997	-0.853	-43.644	2.268	1.475	-5.988	0.367	-0.148	-1.028		1
69	1.134	287.830	-1.502	-33.729	2.292	1.378	-5.551	0.364	-0.161	-0.355	HSR	1
70	1.150	286.127	-1.977	-23.635	2.314	1.290	-4.977	0.362	-0.160	0.556		1
71	1.167	284.053	-2.290	-14.896	2.335	1.212	-4.522	0.359	-0.142	1.427		1

163

Table A.6 (Continued)

DISPLACEMENT/VELOCITY/ACCELERATION DATA GENERATED FROM THE DATA OF THE B. SHANK USING FINITE DIFFERENCE AFTER DIGITAL FILTERING

POS	TIME S	THETA DEG	OMEGA R/S	ALPHA R/S/S	CGX M	VX M/S	AX M/S/S	CGY M	VY M/S	AY M/S/S	MODE1	MODE2
72	1.184	281.752	-2.474	-7.955	2.354	1.139	-4.414	0.357	-0.112	1.930		1
73	1.200	279.328	-2.556	-2.163	2.373	1.065	-4.647	0.355	-0.078	1.891		1
74	1.217	276.870	-2.546	3.075	2.390	0.985	-5.089	0.354	-0.049	1.423		1
75	1.234	274.464	-2.453	7.514	2.405	0.895	-5.605	0.354	-0.031	0.807		1
76	1.250	272.184	-2.296	10.580	2.420	0.798	-5.972	0.353	-0.022	0.280		1
77	1.267	270.079	-2.100	12.132	2.432	0.696	-5.911	0.353	-0.021	-0.033		1
78	1.284	268.172	-1.891	12.549	2.443	0.601	-5.342	0.353	-0.023	-0.114		1
79	1.300	266.467	-1.682	12.320	2.452	0.518	-4.497	0.352	-0.025	-0.040		1
80	1.317	264.959	-1.480	11.756	2.460	0.451	-3.693	0.352	-0.025	0.105	TOI	1
81	1.334	263.639	-1.290	10.979	2.467	0.395	-3.086	0.351	-0.022	0.279		1
82	1.350	262.495	-1.114	9.931	2.473	0.348	-2.672	0.351	-0.015	0.420		1
83	1.367	261.511	-0.959	8.511	2.479	0.306	-2.368	0.351	-0.008	0.453		1
84	1.384	260.663	-0.830	6.917	2.483	0.269	-2.062	0.351	-0.000	0.392		1
85	1.400	259.924	-0.728	5.539	2.488	0.237	-1.661	0.351	-0.006	0.328		1
86	1.417	259.272	-0.646	4.410	2.491	0.213	-1.163	0.351	0.011	0.314		1
87	1.434	258.691	-0.581	3.177	2.495	0.198	-0.668	0.351	0.016	0.320		1
88	1.450	258.162	-0.540	1.640	2.498	0.191	-0.277	0.351	0.021	0.311		1
89	1.467	257.659	-0.527	-0.099	2.501	0.189	0.021	0.352	0.026	0.317		1
90	1.484	257.156	-0.543	-1.921	2.504	0.192	0.335	0.352	0.032	0.382		1
91	1.500	256.622	-0.591	-3.720	2.507	0.200	0.738	0.353	0.039	0.499		1
92	1.517	256.028	-0.667	-5.274	2.511	0.216	1.169	0.354	0.048	0.619		1
93	1.534	255.347	-0.766	-6.303	2.515	0.239	1.541	0.355	0.060	0.702		1
94	1.550	254.564	-0.877	-6.769	2.519	0.268	1.883	0.356	0.072	0.718		1
95	1.567	253.671	-0.992	-7.058	2.524	0.302	2.318	0.357	0.084	0.651		1
96	1.584	252.668	-1.113	-7.667	2.529	0.345	2.972	0.358	0.094	0.553		1
97	1.600	251.546	-1.248	-8.747	2.535	0.401	3.873	0.360	0.102	0.490		1
98	1.617	250.285	-1.404	-9.968	2.542	0.474	4.911	0.362	0.110	0.402		1
99	1.634	248.864	-1.580	-10.932	2.551	0.565	5.937	0.364	0.115	0.179		1
100	1.650	247.267	-1.769	-11.462	2.561	0.672	6.835	0.366	0.116	-0.120		1
101	1.667	245.485	-1.962	-11.371	2.573	0.793	7.444	0.368	0.112	-0.308		1
102	1.684	243.518	-2.148	-10.372	2.588	0.920	7.505	0.369	0.106	-0.288		1
103	1.700	241.382	-2.308	-8.114	2.604	1.043	6.714	0.371	0.102	-0.139		1
104	1.717	239.109	-2.418	-4.097	2.622	1.144	4.805	0.373	0.101	-0.044		1
105	1.734	236.762	-2.445	2.093	2.642	1.203	1.627	0.374	0.100	-0.141		1
106	1.750	234.440	-2.349	10.097	2.662	1.198	-2.634	0.376	0.096	-0.417		1
107	1.767	232.276	-2.108	18.250	2.682	1.115	-7.138	0.378	0.087	-0.705	HSL	1
108	1.784	230.413	-1.740	23.362	2.700	0.960	-10.107	0.379	0.073	-0.800		1
109	1.800	228.952	-1.329	20.731	2.714	0.778	-8.570	0.380	0.060	-0.507		1

Table A.7 Displacement/Velocity/Acceleration—C of G of Thigh

DISPLACEMENT/VELOCITY/ACCELERATION DATA GENERATED FROM THE DATA OF THE R. THIGH USING FINITE DIFFERENCE AFTER DIGITAL FILTERING

PCS	TIME S	THETA DEG	OMEGA R/S	ALPHA R/S/S	CGX M	VX M/S	AX M/S/S	CGY M	VY M/S	AY M/S/S	MODE1	MODE2
1	0.000	294.524	0.018	-1.694	0.544	1.534	11.123	0.752	-0.063	2.190	HST	T
2	0.017	294.528	-0.016	-2.280	0.571	1.694	7.543	0.751	-0.020	2.804		T
3	0.033	294.494	-0.058	-2.990	0.600	1.785	3.344	0.751	0.031	2.940		T
4	0.050	294.417	-0.115	-4.531	0.630	1.805	-0.750	0.752	0.078	2.512		T
5	0.067	294.274	-0.209	-7.390	0.660	1.760	-4.227	0.753	0.114	1.642		T
6	0.083	294.019	-0.362	-11.262	0.689	1.664	-6.690	0.756	0.133	0.655		T
7	0.100	293.584	-0.584	-15.072	0.716	1.537	-7.939	0.758	0.136	-0.074		T
8	0.117	292.903	-0.864	-17.537	0.740	1.399	-8.056	0.760	0.131	-0.360		T
9	0.133	291.933	-1.169	-17.790	0.762	1.268	-7.358	0.762	0.124	-0.315	TOT	T
10	0.150	290.670	-1.457	-15.714	0.783	1.154	-6.238	0.764	0.120	-0.184		T
11	0.167	289.149	-1.693	-11.952	0.801	1.060	-5.044	0.766	0.118	-0.147		T
12	0.183	287.436	-1.856	-7.560	0.818	0.986	-4.004	0.768	0.115	-0.244		T
13	0.200	285.605	-1.945	-3.490	0.834	0.927	-3.218	0.770	0.110	-0.443		T
14	0.217	283.721	-1.972	-0.315	0.849	0.879	-2.656	0.772	0.101	-0.694		T
15	0.233	281.838	-1.955	1.771	0.863	0.838	-2.196	0.773	0.087	-0.970		T
16	0.250	279.986	-1.913	2.915	0.877	0.805	-1.712	0.775	0.068	-1.243		T
17	0.267	278.183	-1.858	3.451	0.890	0.781	-1.146	0.776	0.045	-1.429		T
18	0.283	276.437	-1.798	3.609	0.903	0.767	-0.538	0.776	0.021	-1.413		T
19	0.300	274.749	-1.738	3.438	0.916	0.763	0.015	0.776	-0.002	-1.158		T
20	0.317	273.117	-1.683	2.896	0.928	0.768	0.402	0.776	-0.018	-0.782		T
21	0.333	271.533	-1.641	1.979	0.941	0.777	0.606	0.776	-0.028	-0.457		T
22	0.350	269.982	-1.617	0.900	0.954	0.788	0.758	0.775	-0.033	-0.272		T
23	0.367	268.444	-1.611	0.057	0.967	0.802	1.006	0.775	-0.037	-0.205		T
24	0.383	266.904	-1.615	-0.220	0.981	0.822	1.376	0.774	-0.040	-0.174		T
25	0.400	265.358	-1.619	0.212	0.995	0.848	1.801	0.773	-0.043	-0.118		T
26	0.417	263.812	-1.608	1.362	1.009	0.882	2.236	0.773	-0.044	-0.043		T
27	0.433	262.285	-1.573	3.168	1.024	0.922	2.669	0.772	-0.044	-0.008		T
28	0.450	260.807	-1.503	5.425	1.040	0.971	3.083	0.771	-0.044	-0.053		T
29	0.467	259.415	-1.392	7.789	1.057	1.025	3.479	0.770	-0.046	-0.164		T
30	0.483	258.147	-1.243	10.010	1.074	1.087	3.929	0.770	-0.050	-0.328		T
31	0.500	257.040	-1.059	12.056	1.093	1.156	4.503	0.769	-0.057	-0.589		T

165

Table A.7 (Continued)

DISPLACEMENT/VELOCITY/ACCELERATION DATA GENERATED FROM THE DATA OF THE R. THIGH USING FINITE DIFFERENCE AFTER DIGITAL FILTERING

POS	TIME S	THETA DEG	OMEGA R/S	ALPHA R/S/S	CGX M	VX M/S	AX M/S/S	CGY M	VY M/S	AY M/S/S	EVENT	MODE1	MODE2
32	0.517	256.125	-0.841	13.956	1.113	1.237	5.189	0.768	-0.069	-0.957		1	
33	0.533	255.433	-0.593	15.684	1.134	1.329	5.880	0.766	-0.089	-1.341		1	
34	0.550	254.992	-0.318	17.298	1.157	1.433	6.424	0.765	-0.114	-1.648		1	
35	0.567	254.825	-0.017	18.950	1.182	1.543	6.702	0.763	-0.144	-1.791	HSL	1	
36	0.583	254.960	0.314	20.673	1.208	1.656	6.682	0.760	-0.174	-1.640		1	
37	0.600	255.424	0.673	22.302	1.237	1.766	6.410	0.757	-0.198	-1.128		1	
38	0.617	256.245	1.057	23.602	1.267	1.870	5.915	0.753	-0.211	-0.338		1	
39	0.633	257.444	1.460	24.194	1.299	1.963	5.147	0.750	-0.210	0.612		1	
40	0.650	259.033	1.864	23.380	1.333	2.041	4.002	0.746	-0.191	1.638		1	
41	0.667	261.004	2.239	20.498	1.367	2.097	2.468	0.743	-0.155	2.665		1	
42	0.683	263.310	2.547	15.594	1.403	2.124	-0.765	0.741	-0.102	3.623		1	
43	0.700	265.869	2.759	9.596	1.438	2.122	-0.736	0.740	-0.034	4.447		1	
44	0.717	268.580	2.867	3.623	1.473	2.099	-1.789	0.740	0.046	4.989		1	
45	0.733	271.346	2.880	-1.788	1.508	2.063	-2.431	0.744	0.132	5.049	TOR	1	
46	0.750	274.081	2.807	-6.512	1.542	2.018	-2.839	0.749	0.214	4.523		1	
47	0.767	276.709	2.663	-10.216	1.575	1.968	-3.118	0.754	0.283	3.486		1	
48	0.783	279.168	2.467	-12.468	1.608	1.914	-3.271	0.760	0.331	2.127		1	
49	0.800	281.421	2.247	-13.278	1.639	1.859	-3.283	0.766	0.354	0.686		1	
50	0.817	283.460	2.024	-13.103	1.670	1.805	-3.223	0.772	0.353	-0.616		1	
51	0.833	285.288	1.810	-12.563	1.699	1.752	-3.224	0.777	0.333	-1.683		1	
52	0.850	286.918	1.605	-12.229	1.728	1.697	-3.343	0.781	0.297	-2.546		1	
53	0.867	288.355	1.401	-12.488	1.756	1.640	-3.490	0.785	0.248	-3.223		1	
54	0.884	289.594	1.189	-12.888	1.783	1.581	-3.549	0.788	0.190	-3.645		1	
55	0.900	290.626	0.971	-12.887	1.809	1.522	-3.478	0.789	0.127	-3.759		1	
56	0.917	291.450	0.759	-12.275	1.834	1.465	-3.265	0.790	0.065	-3.637		1	
57	0.934	292.077	0.562	-11.261	1.858	1.413	-2.895	0.790	-0.006	-3.443		1	
58	0.950	292.523	0.384	-9.936	1.881	1.368	-2.393	0.788	-0.050	-3.279		1	
59	0.967	292.810	0.231	-8.203	1.903	1.333	-1.750	0.786	-0.104	-3.092		1	
60	0.984	292.965	0.110	-6.208	1.925	1.310	-0.870	0.783	-0.153	-2.715		1	
61	1.000	293.021	-0.024	-4.322	1.947	1.304	0.299	0.780	-0.194	-2.026		1	
62	1.017	293.010	-0.034	-2.673	1.969	1.320	1.610	0.776	-0.221	-1.080		1	
63	1.034	292.957	-0.065	-1.049	1.991	1.358	2.771	0.772	-0.230	-0.112		1	
64	1.050	292.885	-0.069	0.673	2.014	1.412	3.521	0.768	-0.225	0.637		1	
65	1.067	292.825	-0.043	2.177	2.038	1.475	3.747	0.765	-0.209	1.060		1	
66	1.084	292.804	0.004	2.791	2.063	1.537	3.470	0.762	-0.189	1.229		1	
67	1.100	292.833	0.050	2.001	2.089	1.591	2.794	0.762	-0.168	1.338		1	
68	1.117	292.900	0.071	-0.014	2.116	1.630	1.883	0.757	-0.145	1.558		1	
69	1.134	292.968	-0.050	-2.420	2.144	1.654	0.937	0.757	-0.116	1.946		1	
70	1.150	292.995	-0.010	-4.312	2.171	1.662	0.054	0.755	-0.080	2.406	HSR	1	

Table A.7 (Continued)

DISPLACEMENT/VELOCITY/ACCELERATION DATA GENERATED FROM THE DATA OF THE R. THIGH USING FINITE DIFFERENCE AFTER DIGITAL FILTERING

POS	TIME S	THETA DEG	OMEGA R/S	ALPHA R/S/S	CGX M	VX M/S	AX M/S/S	CGY M	VY M/S	AY M/S/S	MODE1	MODE2
71	1.167	292.949	-0.094	-5.293	2.199	1.656	-0.821	0.754	-0.036	2.710		T
72	1.184	292.815	-0.187	-5.731	2.226	1.634	-1.767	0.754	0.011	2.599		T
73	1.200	292.592	-0.285	-6.489	2.253	1.597	-2.765	0.755	0.051	1.997		T
74	1.217	292.271	-0.403	-8.179	2.280	1.542	-3.721	0.756	0.077	1.150		T
75	1.234	291.823	-0.558	-10.478	2.305	1.473	-4.546	0.757	0.089	0.390		T
76	1.250	291.206	-0.752	-12.190	2.329	1.391	-5.142	0.759	0.090	-0.128		T
77	1.267	290.386	-0.964	-12.265	2.351	1.301	-5.375	0.760	0.085	-0.409		T
78	1.284	289.364	-1.161	-10.775	2.372	1.211	-5.211	0.762	0.077	-0.474		T
79	1.300	288.168	-1.323	-8.672	2.392	1.127	-4.786	0.763	0.069	-0.364		T
80	1.317	286.836	-1.450	-6.837	2.410	1.052	-4.286	0.764	0.064	-0.172	TOT	T
81	1.334	285.397	-1.551	-5.558	2.427	0.984	-3.822	0.765	0.063	-0.001		T
82	1.350	283.872	-1.636	-4.622	2.443	0.924	-3.404	0.766	0.064	0.078		T
83	1.367	282.273	-1.705	-3.643	2.458	0.871	-3.004	0.767	0.066	0.012		T
84	1.384	280.615	-1.757	-2.484	2.472	0.824	-2.595	0.768	0.065	-0.199		T
85	1.400	278.917	-1.788	-1.418	2.485	0.784	-2.149	0.769	0.059	-0.473		T
86	1.417	277.198	-1.804	-0.766	2.498	0.753	-1.624	0.770	0.049	-0.702		T
87	1.434	275.470	-1.814	-0.572	2.510	0.730	-1.027	0.771	0.036	-0.824		T
88	1.450	273.734	-1.823	-0.646	2.522	0.718	-0.426	0.772	0.022	-0.825		T
89	1.467	271.987	-1.835	-0.619	2.534	0.716	-0.125	0.772	0.008	-0.730		T
90	1.484	270.228	-1.844	-0.047	2.546	0.722	0.624	0.772	-0.003	-0.576		T
91	1.500	268.464	-1.837	1.311	2.558	0.737	1.051	0.772	-0.011	-0.387		T
92	1.517	266.719	-1.800	3.198	2.571	0.758	1.350	0.771	-0.016	-0.180		T
93	1.534	265.025	-1.730	4.920	2.583	0.782	1.531	0.771	-0.017	0.000		T
94	1.550	263.414	-1.636	5.937	2.597	0.809	1.725	0.771	-0.016	0.077		T
95	1.567	261.900	-1.532	6.352	2.610	0.839	2.096	0.771	-0.014	-0.002		T
96	1.584	260.487	-1.425	6.698	2.625	0.878	2.744	0.770	-0.016	-0.201		T
97	1.600	259.178	-1.309	7.328	2.640	0.931	3.649	0.770	-0.021	-0.432		T
98	1.617	257.986	-1.180	8.241	2.656	1.000	4.647	0.770	-0.030	-0.666		T
99	1.634	256.924	-1.034	9.494	2.673	1.086	5.512	0.769	-0.043	-0.927		T
100	1.650	256.010	-0.864	11.226	2.692	1.184	6.051	0.768	-0.061	-1.199		T
101	1.667	255.274	-0.660	13.211	2.712	1.288	6.097	0.767	-0.083	-1.394		T
102	1.684	254.750	-0.423	14.768	2.735	1.387	5.506	0.765	-0.108	-1.406	HST	T
103	1.700	254.466	-0.168	15.148	2.759	1.471	4.145	0.763	-0.130	-1.181		T
104	1.717	254.429	0.082	13.989	2.784	1.525	1.797	0.761	-0.147	-0.771		T
105	1.734	254.622	0.299	11.354	2.810	1.531	-1.740	0.759	-0.156	-0.269		T
106	1.750	255.000	0.460	7.586	2.835	1.467	-6.311	0.756	-0.156	0.294		T
107	1.767	255.501	0.552	3.335	2.858	1.321	-11.016	0.753	-0.146	0.910		T
108	1.784	256.054	0.572	-0.108	2.879	1.100	-13.948	0.751	-0.126	1.435		T
109	1.800	256.593	0.548	-0.194	2.895	0.856	-12.037	0.749	-0.098	1.480		T

Table A.8 Displacement/Velocity/Acceleration—C of G of H.A.T.

DISPLACEMENT/VELOCITY/ACCELERATION DATA GENERATED FROM THE DATA OF THE H-A-T   USING FINITE DIFFERENCE AFTER DIGITAL FILTERING

POS	TIME S	THETA DEG	OMEGA R/S	ALPHA R/S/S	CGX M	VX M/S	AX M/S/S	CGY M	VY M/S	AY M/S/S	MODE1	MODE2
1	0.000	97.747	0.280	0.388	0.426	1.442	11.419	1.242	-0.088	0.504	HSR	1
2	0.017	98.018	0.282	-0.224	0.452	1.610	8.203	1.241	-0.077	0.875		1
3	0.033	98.286	0.272	-1.076	0.480	1.715	4.380	1.239	-0.058	1.419		1
4	0.050	98.538	0.246	-2.099	0.509	1.756	0.734	1.239	-0.029	2.182		1
5	0.067	98.757	0.202	-3.181	0.539	1.740	-2.131	1.238	0.014	2.941		1
6	0.083	98.924	0.140	-4.133	0.567	1.685	-3.816	1.239	0.069	3.355		1
7	0.100	99.025	0.064	-4.858	0.595	1.613	-4.249	1.241	0.126	3.294		1
8	0.117	99.047	-0.022	-5.453	0.621	1.543	-4.693	1.243	0.179	2.909		1
9	0.133	98.983	-0.117	-6.101	0.646	1.489	-2.596	1.247	0.223	2.378		1
10	0.150	98.823	-0.225	-6.877	0.671	1.457	-1.415	1.251	0.258	1.797	TOT	1
11	0.167	98.553	-0.347	-7.641	0.695	1.442	-0.473	1.255	0.283	1.249		1
12	0.183	98.161	-0.480	-8.128	0.719	1.441	0.082	1.260	0.300	0.746		1
13	0.200	97.637	-0.618	-8.076	0.743	1.445	0.242	1.265	0.308	0.180		1
14	0.217	96.981	-0.749	-7.299	0.767	1.449	0.086	1.270	0.306	-0.587		1
15	0.233	96.206	-0.861	-5.800	0.791	1.441	-0.233	1.275	0.288	-1.594		1
16	0.250	95.336	-0.942	-3.808	0.815	1.441	-0.562	1.280	0.252	-2.701		1
17	0.267	94.405	-0.988	-1.604	0.839	1.429	-0.836	1.284	0.198	-3.590		1
18	0.283	93.449	-0.996	0.552	0.863	1.413	-1.047	1.287	0.133	-3.948		1
19	0.300	92.503	-0.970	2.330	0.886	1.394	-1.169	1.288	0.067	-3.737		1
20	0.317	91.597	-0.918	3.512	0.909	1.374	-1.200	1.289	0.008	-3.219		1
21	0.333	90.749	-0.852	4.202	0.932	1.354	-1.167	1.289	-0.041	-2.679		1
22	0.350	89.968	-0.778	4.558	0.954	1.335	-1.023	1.288	-0.081	-2.205		1
23	0.367	89.262	-0.700	4.485	0.977	1.320	-0.648	1.286	-0.114	-1.757		1
24	0.383	88.630	-0.629	3.922	0.998	1.314	-0.057	1.284	-0.140	-1.326		1
25	0.400	88.062	-0.570	3.150	1.020	1.318	0.550	1.281	-0.158	-0.974		1
26	0.417	87.542	-0.524	2.519	1.042	1.332	0.999	1.278	-0.172	-0.752		1
27	0.433	87.062	-0.486	2.093	1.065	1.352	1.259	1.275	-0.183	-0.654		1
28	0.450	86.614	-0.454	1.902	1.087	1.374	1.340	1.272	-0.194	-0.595		1
29	0.467	86.195	-0.422	2.219	1.111	1.396	1.225	1.269	-0.203	-0.465		1
30	0.483	85.807	-0.380	3.150	1.134	1.415	1.016	1.266	-0.209	-0.240		1
31	0.500	85.469	-0.317	4.225	1.158	1.436	0.954	1.262	-0.211	0.028		1

Table A.8 (Continued)

DISPLACEMENT/VELOCITY/ACCELERATION DATA GENERATED FROM THE DATA OF THE H-A-T USING FINITE DIFFERENCE AFTER DIGITAL FILTERING

POS	TIME S	THETA DEG	OMEGA R/S	ALPHA R/S/S	CGX M	VX M/S	AX M/S/S	CGY M	VY M/S	AY M/S/S	MODE1	MODE2
32	0.517	85.201	-0.239	4.834	1.182	1.447	1.212	1.258	-0.209	0.317		1
33	0.533	85.012	-0.156	4.835	1.206	1.470	1.697	1.255	-0.201	0.610		1
34	0.550	84.903	-0.076	4.732	1.231	1.503	2.104	1.252	-0.188	0.861		1
35	0.567	84.867	0.002	4.892	1.256	1.541	2.100	1.249	-0.172	1.096		1
36	0.583	84.906	0.087	5.846	1.282	1.573	1.491	1.246	-0.152	1.417		1
37	0.600	85.033	0.196	7.535	1.309	1.590	0.361	1.244	-0.125	1.829	HSL	1
38	0.617	85.261	0.338	9.083	1.335	1.585	-0.941	1.242	-0.091	2.188		1
39	0.633	85.679	0.499	9.443	1.361	1.559	-2.051	1.241	-0.052	2.386		1
40	0.650	86.235	0.653	8.312	1.387	1.517	-2.821	1.240	-0.011	2.449		1
41	0.667	86.927	0.776	6.289	1.412	1.465	-3.307	1.240	0.030	2.434		1
42	0.683	87.718	0.863	4.137	1.436	1.407	-3.520	1.241	0.070	2.409		1
43	0.700	88.575	0.914	2.092	1.459	1.348	-3.314	1.243	0.110	2.444		1
44	0.717	89.465	0.932	0.127	1.481	1.296	-2.633	1.245	0.152	2.502		1
45	0.733	90.356	0.919	-1.536	1.502	1.260	-1.713	1.248	0.194	2.460	TOP	1
46	0.750	91.220	0.881	-2.651	1.523	1.239	-0.867	1.251	0.234	2.232		1
47	0.767	92.039	0.830	-3.256	1.543	1.231	-0.260	1.256	0.268	1.770		1
48	0.783	92.805	0.773	-3.528	1.564	1.230	0.088	1.260	0.293	1.036		1
49	0.800	93.515	0.713	-3.694	1.584	1.234	0.281	1.265	0.302	-0.047		1
50	0.817	94.167	0.650	-4.054	1.605	1.240	0.447	1.270	0.294	-1.057		1
51	0.833	94.756	0.577	-4.787	1.626	1.249	0.617	1.275	0.267	-2.063		1
52	0.850	95.270	0.490	-5.722	1.647	1.260	0.764	1.279	0.225	-2.811		1
53	0.867	95.692	0.387	-6.517	1.668	1.274	0.886	1.283	0.174	-3.197		1
54	0.884	96.008	0.273	-7.034	1.689	1.290	0.992	1.285	0.119	-3.168		1
55	0.900	96.212	0.152	-7.227	1.711	1.307	1.034	1.287	0.068	-2.818		1
56	0.917	96.299	0.032	-6.838	1.733	1.324	0.949	1.287	-0.025	-2.410		1
57	0.934	96.273	-0.076	-5.682	1.755	1.339	0.754	1.287	-0.012	-2.198		1
58	0.950	96.154	-0.158	-3.997	1.777	1.350	0.526	1.286	-0.049	-2.245		1
59	0.967	95.972	-0.209	-2.008	1.800	1.356	0.308	1.284	-0.087	-2.410		1
60	0.984	95.755	-0.225	0.490	1.823	1.360	0.116	1.281	-0.129	-2.499		1
61	1.000	95.542	-0.193	3.482	1.845	1.360	-0.032	1.278	-0.171	-2.346		1
62	1.017	95.387	-0.109	6.230	1.868	1.365	0.154	1.278	-0.207	-1.819		1
63	1.034	95.335	0.015	7.687	1.891	1.361	0.448	1.275	-0.231	-0.950		1
64	1.050	95.415	0.154	8.088	1.913	1.376	0.792	1.267	-0.239	0.016		1
65	1.067	95.630	0.285	6.863	1.936	1.392	1.129	1.267	-0.231	0.794		1
66	1.084	95.959	0.383	4.545	1.960	1.413	1.483	1.263	-0.212	1.342		1
67	1.100	96.362	0.436	1.839	1.984	1.474	1.821	1.260	-0.190	1.245	HSP	1
68	1.117	96.792	0.444	-0.494	2.008	1.474	2.045	1.257	-0.168	1.245		1
69	1.134	97.211	0.420	-2.069	2.033	1.509	2.080	1.254	-0.148	1.148		1
70	1.150	97.593	0.375	-2.924	2.058	1.543	1.853	1.252	-0.129	1.254		1

Table A.8 (Continued)

DISPLACEMENT/VELOCITY/ACCELERATION DATA GENERATED FROM THE DATA OF THE H-A-T USING FINITE DIFFERENCE AFTER DIGITAL FILTERING

POS	TIME S	THETA DEG	OMEGA R/S	ALPHA R/S/S	CGX M	VX M/S	AX M/S/S	CGY M	VY M/S	AY M/S/S	MODE1	MODE2
71	1.167	97.928	0.322	-3.302	2.084	1.571	1.288	1.250	-0.106	1.622	1	1
72	1.184	98.209	0.265	-3.505	2.111	1.586	0.442	1.248	-0.075	2.112	1	1
73	1.200	98.435	0.205	-3.772	2.137	1.586	-0.439	1.247	-0.036	2.535	1	1
74	1.217	98.601	0.140	-4.174	2.163	1.572	-1.109	1.247	0.009	2.788	1	1
75	1.234	98.701	0.066	-4.751	2.190	1.549	-1.467	1.247	0.057	2.859	1	1
76	1.250	98.727	-0.015	-5.733	2.215	1.523	-1.496	1.249	0.105	2.767	1	1
77	1.267	98.665	-0.125	-7.230	2.240	1.499	-1.201	1.251	0.149	2.522	1	1
78	1.284	98.489	-0.260	-8.670	2.265	1.483	-0.724	1.254	0.189	2.100	1	1
79	1.300	98.169	-0.414	-8.968	2.290	1.475	-0.414	1.257	0.219	1.496	1	1
80	1.317	97.698	-0.559	-7.538	2.314	1.469	-0.547	1.261	0.239	0.800	TOI	1
81	1.334	97.101	-0.665	-4.974	2.337	1.457	-1.010	1.265	0.246	0.173	1	1
82	1.350	96.427	-0.725	-2.505	2.363	1.435	-1.519	1.269	0.244	-0.316	1	1
83	1.367	95.717	-0.749	-0.992	2.387	1.406	-1.750	1.273	0.236	-0.782	1	1
84	1.384	94.996	-0.758	-0.508	2.410	1.376	-1.686	1.277	0.218	-1.359	1	1
85	1.400	94.269	-0.766	-0.556	2.432	1.350	-1.417	1.281	0.190	-2.012	1	1
86	1.417	93.534	-0.776	-0.639	2.455	1.329	-1.022	1.283	0.151	-2.588	1	1
87	1.434	92.787	-0.787	-0.539	2.477	1.316	-0.567	1.286	0.104	-2.968	1	1
88	1.450	92.030	-0.794	-0.197	2.499	1.310	-0.160	1.287	0.052	-3.096	1	1
89	1.467	91.269	-0.794	0.371	2.520	1.310	0.106	1.287	0.001	-2.995	1	1
90	1.484	90.514	-0.782	1.036	2.542	1.314	0.183	1.287	-0.048	-2.747	1	1
91	1.500	89.776	-0.757	1.791	2.564	1.317	0.065	1.286	-0.091	-2.412	1	1
92	1.517	89.067	-0.722	2.307	2.586	1.316	-0.191	1.284	-0.128	-2.005	1	1
93	1.534	88.396	-0.681	2.628	2.608	1.310	-0.451	1.282	-0.158	-1.540	1	1
94	1.550	87.767	-0.635	2.962	2.630	1.301	-0.550	1.279	-0.179	-1.054	1	1
95	1.567	87.184	-0.582	3.506	2.651	1.292	-0.403	1.276	-0.193	-0.611	1	1
96	1.584	86.655	-0.518	4.258	2.673	1.287	-0.006	1.272	-0.200	-0.240	1	1
97	1.600	86.195	-0.440	5.109	2.694	1.292	0.585	1.269	-0.201	-0.088	1	1
98	1.617	85.815	-0.347	5.895	2.716	1.307	1.249	1.266	-0.197	0.352	1	1
99	1.634	85.531	-0.243	6.282	2.738	1.333	1.855	1.262	-0.189	0.471	1	1
100	1.650	85.350	-0.138	5.993	2.760	1.369	2.262	1.259	-0.181	0.447	1	1
101	1.667	85.268	-0.044	5.096	2.784	1.409	2.293	1.256	-0.174	0.382	1	1
102	1.684	85.267	-0.032	4.054	2.807	1.445	1.769	1.253	-0.168	0.393	1	1
103	1.700	85.329	0.092	3.345	2.832	1.468	0.547	1.251	-0.161	0.532	1	1
104	1.717	85.442	0.144	2.990	2.856	1.463	-1.512	1.248	-0.151	0.744	1	1
105	1.734	85.603	0.191	2.623	2.881	1.417	-4.500	1.246	-0.136	0.965	HST	1
106	1.750	85.808	0.231	1.781	2.904	1.313	-8.170	1.244	-0.118	1.223	1	1
107	1.767	86.044	0.251	0.312	2.924	1.145	-11.687	1.242	-0.096	1.511	1	1
108	1.784	86.287	0.241	-1.179	2.942	0.924	-13.554	1.240	-0.068	1.653	1	1
109	1.800	86.506	0.211	-1.355	2.955	0.693	-11.570	1.239	-0.040	1.455	1	1

Table A.9 Displacement/Velocity/Acceleration—Total Body C of G

OUTPUT OF TOTAL BODY CENTRE OF GRAVITY DATA FOR EACH POSITION

POSITION	TIME s	TBCGX M	TBCGY M	TBVX M/S	TBVY M/S	TBAX M/S/S	TBAY M/S/S	TBPE JOULES	TBKE JOULES	TBE JOULES	SUMDE JOULES
1	0.00000	0.446	1.031	1.426	-0.090	10.373	0.429	808.142	81.579	889.721	0.000
2	0.01667	0.471	1.029	1.580	-0.079	7.704	0.914	807.010	100.050	907.060	17.339
3	0.03334	0.498	1.028	1.683	-0.060	4.516	1.472	806.074	113.312	919.385	79.664
4	0.05001	0.527	1.027	1.731	-0.030	1.431	2.094	805.451	119.762	925.213	35.492
5	0.06668	0.556	1.027	1.730	0.010	-1.093	2.619	805.289	119.680	924.968	35.737
6	0.08335	0.585	1.028	1.694	0.057	-2.744	2.842	805.716	114.868	920.584	40.121
7	0.10002	0.613	1.029	1.639	0.105	-3.423	2.724	806.786	107.796	914.582	46.123
8	0.11669	0.639	1.031	1.580	0.148	-3.269	2.407	808.458	100.675	909.133	51.572
9	0.13336	0.665	1.034	1.530	0.185	-2.588	1.998	810.658	94.923	905.580	55.125
10	0.15003	0.690	1.037	1.494	0.215	-1.742	1.526	813.297	91.040	904.336	56.369
11	0.16670	0.715	1.041	1.472	0.236	-1.017	1.027	816.271	88.814	905.085	57.117
12	0.18337	0.740	1.045	1.460	0.249	-0.558	0.519	819.466	87.673	907.139	59.172
13	0.20004	0.764	1.049	1.453	0.247	-0.383	-0.046	822.778	86.978	909.756	61.788
14	0.21671	0.788	1.054	1.447	0.229	-0.430	-0.728	826.088	86.160	912.248	64.280
15	0.23338	0.812	1.058	1.439	0.229	-0.594	-1.528	829.246	84.852	914.098	66.130
16	0.25005	0.836	1.061	1.427	0.196	-0.781	-2.350	832.074	82.982	915.056	67.088
17	0.26672	0.860	1.064	1.413	0.151	-0.950	-2.985	834.382	80.697	915.078	67.110
18	0.28339	0.883	1.066	1.396	0.097	-1.081	-3.215	836.013	78.242	914.255	67.934
19	0.30006	0.906	1.067	1.377	0.044	-1.154	-3.010	836.916	75.845	912.762	69.427
20	0.31673	0.929	1.068	1.357	-0.003	-1.179	-2.554	837.150	73.632	910.782	71.406
21	0.33340	0.951	1.067	1.337	-0.042	-1.180	-2.069	836.828	71.574	908.402	73.786
22	0.35007	0.974	1.066	1.318	-0.072	-1.104	-1.653	836.062	69.635	905.696	76.492
23	0.36674	0.995	1.065	1.301	-0.097	-0.849	-1.293	834.937	67.997	902.934	79.254
24	0.38341	1.017	1.063	1.290	-0.129	-0.415	-0.966	833.532	67.006	900.538	81.650
25	0.40008	1.038	1.061	1.287	-0.129	0.072	-0.691	831.919	66.858	898.777	83.411
26	0.41675	1.060	1.059	1.292	-0.138	0.500	-0.490	830.161	67.490	897.651	84.538
27	0.43342	1.081	1.057	1.304	-0.145	0.841	-0.385	828.300	68.765	897.065	85.124
28	0.45009	1.103	1.054	1.320	-0.151	1.063	-0.334	826.360	70.566	896.925	85.263
29	0.46676	1.125	1.051	1.339	-0.157	1.079	-0.273	824.345	72.643	896.987	85.325
30	0.48343	1.148	1.049	1.357	-0.160	1.079	-0.184	822.268	74.677	896.945	85.368
31	0.50010	1.171	1.046	1.375	-0.163	1.122	-0.084	820.152	76.625	896.777	85.536
32	0.51672	1.194	1.043	1.395	-0.163	1.379	0.055	818.015	78.829	896.844	85.603
33	0.53344	1.217	1.041	1.421	-0.161	1.786	0.253	815.885	81.738	897.624	86.383
34	0.55011	1.241	1.038	1.454	-0.155	2.133	0.493	813.810	85.505	899.314	88.073

Table A.9 (Continued)

OUTPUT OF TOTAL BODY CENTRE OF GRAVITY DATA FOR EACH POSITION

POSITION	TIME	TBCGX	TBCGY	TBVX	TBVY	TBAX	TBAY	TBPE	TBKE	TBE	SUMDE
	S	M	M	M/S	M/S	M/S/S	M/S/S	JOULES	JOULES	JOULES	JOULES
35	0.56678	1.266	1.036	1.492	-0.144	2.184	0.782	811.840	89.817	901.657	90.417
36	0.58345	1.291	1.033	1.527	-0.129	1.788	1.155	810.035	93.888	903.923	92.682
37	0.60012	1.317	1.031	1.554	-0.106	0.989	1.569	808.476	96.684	905.160	93.920
38	0.61679	1.343	1.030	1.560	-0.076	0.032	1.894	807.266	97.529	904.795	94.285
39	0.63346	1.369	1.029	1.553	-0.043	-0.835	2.061	806.480	96.441	902.920	96.074
40	0.65013	1.394	1.028	1.532	-0.008	-1.528	2.112	806.148	93.858	900.006	99.074
41	0.66680	1.420	1.028	1.502	0.028	-2.084	2.098	806.279	90.178	896.457	102.623
42	0.68347	1.444	1.029	1.463	0.062	-2.460	2.064	806.870	85.693	892.562	106.518
43	0.70014	1.468	1.030	1.420	0.096	-2.494	2.064	807.907	80.943	888.849	110.231
44	0.71681	1.492	1.032	1.380	0.131	-2.135	2.069	809.391	76.776	886.167	112.913
45	0.73348	1.514	1.035	1.349	0.165	-1.575	1.974	811.333	73.790	886.123	113.957
46	0.75015	1.537	1.038	1.327	0.197	-1.055	1.710	813.716	71.957	885.673	114.507
47	0.76682	1.559	1.041	1.313	0.222	-0.692	1.253	816.481	70.930	887.411	116.245
48	0.78349	1.581	1.045	1.304	0.239	-0.484	0.589	819.531	70.259	889.790	118.624
49	0.80016	1.602	1.049	1.297	0.242	-0.341	-0.239	822.719	69.611	892.330	121.164
50	0.81683	1.624	1.053	1.293	0.231	-0.187	-1.103	825.859	68.928	894.787	123.621
51	0.83350	1.645	1.057	1.291	0.205	-0.023	-1.846	828.750	68.308	897.058	125.892
52	0.85017	1.667	1.060	1.292	0.169	0.111	-2.370	831.226	67.865	899.090	127.924
53	0.86684	1.688	1.063	1.295	0.126	0.208	-2.625	833.172	67.642	900.814	129.648
54	0.88351	1.710	1.064	1.299	0.082	0.293	-2.570	834.527	67.707	902.234	131.068
55	0.90018	1.732	1.065	1.304	0.041	0.340	-2.258	835.306	68.084	903.390	132.224
56	0.91685	1.753	1.066	1.310	0.006	0.306	-1.882	835.590	68.624	904.213	133.047
57	0.93352	1.775	1.066	1.315	-0.022	0.196	-1.640	835.473	69.107	904.580	133.414
58	0.95019	1.797	1.065	1.317	-0.048	-0.051	-1.596	835.013	69.400	904.413	133.581
59	0.96686	1.819	1.064	1.316	-0.075	-0.075	-1.667	834.211	69.494	903.705	134.288
60	0.98353	1.841	1.063	1.314	-0.104	-0.128	-1.713	833.245	69.476	902.521	135.472
61	1.00020	1.863	1.061	1.312	-0.132	-0.029	-1.604	831.496	69.521	901.017	136.976
62	1.01687	1.885	1.058	1.313	-0.157	-0.273	-1.243	829.584	69.932	899.516	138.477
63	1.03354	1.907	1.055	1.321	-0.174	0.705	-0.667	827.384	70.985	898.369	139.624
64	1.05021	1.929	1.052	1.337	-0.180	1.128	-0.063	825.039	72.722	897.761	140.232
65	1.06688	1.951	1.049	1.359	-0.176	1.471	0.384	822.691	75.045	897.736	140.258
66	1.08355	1.974	1.046	1.386	-0.167	1.731	0.618	820.440	77.884	898.323	140.845
67	1.10021	1.998	1.044	1.417	-0.155	1.867	0.687	818.333	81.176	899.508	142.030
68	1.11688	2.022	1.041	1.448	-0.144	1.799	0.703	816.378	84.650	901.028	143.550
69	1.13355	2.046	1.039	1.477	-0.132	1.463	0.785	814.573	87.843	902.417	144.938
70	1.15022	2.071	1.037	1.497	-0.118	0.777	1.027	812.929	90.118	903.047	145.569
71	1.16689	2.096	1.035	1.502	-0.098	-0.305	1.423	811.498	90.615	902.113	146.503
72	1.18356	2.121	1.034	1.487	-0.076	-1.263	1.845	810.375	88.547	898.922	149.694
73	1.20023	2.145	1.033	1.448	-0.036	-2.763	2.146	809.663	83.859	898.522	155.094
74	1.21690	2.169	1.032	1.395	0.001	-3.041	2.256	809.429	77.742	887.171	161.444

Table A.10 Force Plate Data—Forces, Moment and Center of Pressure

M_z	F_x	F_y	X_{cp}	Y_{cp}
0.27	15.42	18.51	76.39	0.00
26.55	-28.61	201.62	77.13	0.00
53.96	-79.14	387.38	77.86	0.00
79.14	-131.09	557.42	78.60	0.00
89.56	-172.15	689.04	80.51	0.00
99.19	-195.26	775.69	81.18	0.00
100.68	-202.19	829.94	82.09	0.00
98.04	-197.72	866.88	83.00	0.00
90.70	-185.82	890.29	84.09	0.00
84.22	-169.30	895.56	84.77	0.00
73.85	-149.96	878.73	85.63	0.00
60.83	-129.13	840.91	86.65	0.00
48.31	-108.33	788.51	87.62	0.00
34.11	-89.35	731.16	88.97	0.00
18.87	-73.73	677.98	90.78	0.00
9.53	-61.76	634.74	92.03	0.00
0.09	-52.25	603.90	93.51	0.00
-7.33	-43.87	585.63	94.78	0.00
-15.19	-36.06	578.60	96.16	0.00
-18.97	-28.59	580.96	96.81	0.00
-23.79	-21.15	591.01	97.58	0.00
-27.17	-13.37	607.13	98.03	0.00
-30.88	-4.89	627.66	98.46	0.00
-35.63	4.50	651.10	98.98	0.00
-37.25	14.98	676.18	98.97	0.00
-43.55	26.91	702.24	99.60	0.00
-47.83	40.58	729.17	99.88	0.00
-55.48	55.86	756.84	100.57	0.00
-58.65	72.39	784.09	100.65	0.00
-63.38	89.60	808.28	100.97	0.00
-70.02	106.64	825.89	101.60	0.00
-72.44	122.73	833.53	101.85	0.00
-75.43	137.45	828.18	102.35	0.00
-77.11	150.32	806.79	102.93	0.00
-74.35	160.25	765.96	103.25	0.00
-71.18	165.14	702.53	103.88	0.00
-63.93	161.94	615.12	104.39	0.00
-53.25	147.83	506.43	104.85	0.00
-42.54	122.43	385.29	105.89	0.00
-29.38	89.41	265.59	106.74	0.00
-16.16	55.50	161.54	107.00	0.00
-6.77	27.22	82.11	107.29	0.00
-1.82	7.34	27.24	109.10	0.00
1.11	-6.38	-12.01	110.90	0.00

Table A.11 Joint Angles, Velocities and Accelerations

FRAME	ANGLE (DEGREES)			ANGULAR VELOCITY (R/S)			ANGULAR ACCELERATION (R/S/S)		
	ANKLE	KNEE	HIP	ANKLE	KNEE	HIP	ANKLE	KNEE	HIP
1	-1.2	5.4	24.5	1.21	2.01	0.02	23.62	19.95	-1.71
2	-0.2	7.5	24.5	1.53	2.29	-0.02	10.18	14.49	-2.18
3	0.7	9.8	24.5	1.55	2.49	-0.06	-9.41	9.07	-2.73
4	3.1	12.2	24.4	1.22	2.59	0.11	-27.19	2.41	-4.18
5	4.1	14.7	24.3	0.65	2.57	-0.20	-36.55	-6.61	-7.11
6	4.4	17.1	24.1	0.00	2.37	-0.35	-36.64	-17.09	-11.22
7	4.4	19.3	23.6	-0.58	2.00	-0.57	-30.56	-26.55	-15.33
8	3.3	20.9	23.0	-1.02	1.49	-0.86	-21.45	-32.67	-18.01
9	2.1	22.1	22.0	-1.29	0.91	-1.17	-11.57	-36.34	-18.31
10	0.8	22.7	20.7	-1.40	0.34	-1.47	-2.95	-31.81	-16.10
11	-0.6	22.7	19.2	-1.39	-0.15	-1.71	3.26	-26.46	-12.11
12	-1.9	22.4	17.5	-1.29	-0.54	-1.87	7.22	-20.08	-7.48
13	-3.1	21.6	15.6	-1.15	-0.82	-1.96	9.56	-13.98	-3.25
14	-4.1	20.8	13.7	-0.98	-1.01	-1.98	10.37	-8.81	-0.03
15	-4.9	19.8	11.8	-0.80	-1.11	-1.96	9.47	-4.71	2.01
16	-5.6	18.7	10.0	-0.66	-1.16	-1.91	7.23	-1.49	3.08
17	-6.2	17.6	8.2	-0.56	-1.16	-1.86	4.72	1.09	3.54
18	-6.7	16.5	6.4	-0.46	-1.13	-1.79	2.92	3.02	3.67
19	-7.1	15.4	4.8	-0.44	-1.06	-1.73	1.84	3.24	3.49
20	-7.5	14.4	3.1	-0.43	-0.99	-1.68	1.02	4.72	2.95
21	-8.0	13.5	1.6	-0.43	-0.91	-1.63	0.41	4.43	2.00
22	-8.4	12.7	0.0	-0.43	-0.84	-1.61	0.21	3.66	0.85
23	-8.8	11.9	-1.5	-0.43	-0.78	-1.61	-0.48	3.10	-0.05
24	-9.2	11.2	-3.1	-0.42	-0.73	-1.61	1.27	3.44	-0.37
25	-9.6	10.5	-4.6	-0.38	-0.67	-1.62	2.46	4.94	-0.96
26	-9.9	9.9	-6.2	-0.33	-0.57	-1.61	3.53	7.30	1.24
27	-10.2	9.9	-7.7	-0.26	-0.43	-1.58	4.06	10.09	3.10
28	-10.4	9.1	-9.2	-0.19	-0.23	-1.51	4.30	13.06	5.44
29	-10.6	9.0	-10.6	-0.12	0.01	-1.40	4.95	16.06	7.85
30	-10.6	9.1	-11.8	-0.03	0.30	-1.25	6.54	19.04	10.28
31	-10.6	9.6	-12.9	0.10	0.64	-1.06	9.37	22.02	12.11

Table A.11 *(Continued)*

FRAME	ANGLE (DEGREES)			ANGULAR VELOCITY (R/S)			ANGULAR ACCELERATION (R/S/S)		
	ANKLE	KNEE	HIP	ANKLE	KNEE	HIP	ANKLE	KNEE	HIP
32	-10.4	10.4	-13.9	0.28	1.04	-0.84	13.90	24.88	13.96
33	-10.1	11.6	-14.5	0.56	1.47	-0.59	20.49	27.57	15.63
34	-9.4	13.2	-15.0	0.97	1.96	-0.32	28.61	30.14	17.19
35	-8.2	15.9	-15.2	1.52	2.48	-0.02	36.51	32.41	18.81
36	-6.4	17.9	-15.1	2.18	3.04	0.31	41.34	33.62	20.56
37	-4.0	21.1	-14.6	2.89	3.60	0.66	39.51	32.96	22.27
38	-1.1	24.8	-13.8	3.50	4.13	1.05	28.05	30.26	23.71
39	2.6	29.0	-12.6	3.83	4.61	1.45	6.73	25.48	24.47
40	6.4	33.6	-11.0	3.73	4.98	1.86	-20.77	17.69	23.78
41	9.8	38.5	-9.0	3.14	5.18	2.25	-47.60	5.75	20.88
42	12.8	43.5	-6.7	2.14	5.10	2.56	-66.70	-9.88	15.81
43	13.8	48.5	-4.1	0.91	4.87	2.77	-73.96	-26.67	9.61
44	14.1	52.8	-1.4	-0.33	4.29	2.88	-69.35	-41.58	3.49
45	13.2	56.6	1.4	-1.40	3.48	2.89	-55.67	-52.91	-2.01
46	11.4	59.5	4.1	-2.18	2.52	2.81	-37.38	-60.21	-6.79
47	9.0	61.8	6.8	-2.65	1.47	2.66	-19.17	-63.27	-10.52
48	6.7	62.3	9.2	-2.82	0.41	2.86	-4.08	-62.00	-12.73
49	3.7	62.2	11.5	-2.78	-0.59	2.24	7.16	-57.06	-13.41
50	1.3	61.2	13.5	-2.58	-1.49	2.02	15.17	-49.82	-13.08
51	-1.3	59.3	15.3	-2.28	-2.25	1.80	20.96	-41.89	-12.41
52	-3.9	56.8	16.9	-1.88	-2.89	1.60	25.83	-34.83	-12.07
53	-6.0	53.8	18.4	-1.41	-3.41	1.40	30.75	-29.59	-12.35
54	-6.5	50.3	19.6	-0.86	-3.87	1.19	35.37	-25.81	-12.86
55	-6.1	46.2	20.7	-0.41	-4.28	0.97	38.00	-22.24	-12.02
56	-5.7	42.2	21.5	0.98	-4.61	0.76	36.50	-17.96	-12.34
57	-4.1	37.6	22.1	1.38	-4.87	0.56	29.25	-12.75	-11.34
58	-3.6	32.4	22.6	1.52	-5.04	0.38	16.26	-6.49	-10.02
59	-1.5	28.0	22.8	1.38	-5.09	0.23	-0.15	1.23	-8.76
60	-0.5	23.1	23.0	0.49	-5.00	0.11	-15.91	10.68	-6.22
61	-0.2	18.4	23.0	0.00	-4.73	0.02	-26.72	21.53	-6.34
62	-0.1	14.1	23.0	-0.49	-4.28	0.04	-29.91	33.00	-2.77
63	-0.2	10.1	23.0	-0.37	-3.63	0.07	-25.81	43.98	-1.06
64	-0.9	7.1	22.9	-0.57	-2.81	0.05	-16.95	53.15	0.73
65	-1.6	4.9	22.8	-0.58	-1.86	0.06	-6.06	58.94	3.04
66	-1.6	3.6	22.8	-0.40	-0.85	0.08	5.04	59.66	2.19
67	-1.6	3.3	22.8	-0.08	0.13	0.06	15.02	54.34	2.02
68	-1.6	3.8	22.9	0.34	0.96	0.04	22.14	43.90	-2.54
69	-1.5	5.1	23.0	0.34	1.59	0.04	24.14	30.98	-2.54
70	-1.0	6.9	23.0	0.73	2.00	-0.01	19.57	18.65	-4.49

175

Table A.11 (Continued)

FRAME	ANGLE (DEGREES)			ANGULAR VELOCITY (R/S)			ANGULAR ACCELERATION (R/S/S)		
	ANKLE	KNEE	HIP	ANKLE	KNEE	HIP	ANKLE	KNEE	HIP
71	-0.1	8.9	23.0	0.99	2.21	-0.09	8.34	8.97	-5.38
72	-0.9	11.1	22.8	1.01	2.29	-0.19	-7.06	1.91	-5.65
73	1.8	13.3	22.6	0.75	2.28	-0.28	-21.31	4.34	-6.31
74	2.3	15.4	22.3	0.30	2.15	-0.40	-29.61	-11.25	-8.07
75	2.4	17.4	21.9	-0.23	1.90	-0.55	-30.36	-18.20	-10.57
76	1.0	19.2	21.2	-0.71	1.54	-0.71	-25.39	-23.16	-12.46
77	-0.2	20.8	20.4	-1.08	1.13	-0.97	-17.42	-24.75	-12.55
78	-0.5	21.2	19.4	-1.30	0.72	-1.17	-8.64	-23.47	-10.91
79	-1.1	21.9	18.2	-1.37	0.35	-1.33	-0.62	-20.92	-8.64
80	-2.8	21.8	16.9	-1.32	0.02	-1.46	5.63	-18.41	-6.71
81	-2.0	21.8	15.4	-1.18	-0.27	-1.54	9.59	-16.38	-5.45
82	-5.0	21.4	13.9	-1.00	-0.53	-1.64	10.99	-14.48	-4.58
83	-5.9	20.8	12.3	-0.81	-0.75	-1.71	10.06	-12.09	-3.62
84	-6.6	19.0	10.9	-0.66	-0.03	-1.76	7.96	-9.28	-2.44
85	-7.0	17.9	8.9	-0.55	-1.06	-1.79	5.95	-6.82	-1.34
86	-7.6	16.8	7.2	-0.46	-1.16	-1.80	2.73	-5.08	-0.68
87	-8.0	15.6	5.4	-0.40	-1.28	-1.81	2.28	-3.75	-0.53
88	-8.4	14.7	3.8	-0.37	-1.31	-1.82	1.53	-2.38	-0.69
89	-8.7	13.7	2.0	-0.35	-1.30	-1.83	1.26	-0.67	-0.74
90	-9.4	11.8	0.3	-0.33	-1.25	-1.84	1.82	0.74	-0.20
91	-9.6	10.7	-1.5	-0.29	-1.14	-1.84	2.35	5.01	0.23
92	-9.8	9.7	-3.3	-0.25	-0.96	-1.80	2.29	8.62	1.23
93	-9.8	8.8	-5.0	-0.22	-0.76	-1.73	2.05	11.44	3.26
94	-10.0	8.2	-6.6	-0.18	-0.54	-1.63	2.27	12.80	5.07
95	-10.3	7.6	-8.1	-0.09	-0.31	-1.53	2.90	13.28	6.06
96	-10.4	7.7	-9.5	-0.02	-0.07	-1.42	3.56	14.08	6.38
97	-10.3	8.0	-10.8	0.05	0.21	-1.31	4.13	15.70	6.62
98	-10.1	8.7	-13.1	0.14	0.53	-1.18	4.86	17.82	7.15
99	-10.1	9.7	-14.0	0.26	0.89	-1.04	6.26	20.15	8.01
100	-9.8	11.1	-14.7	0.68	1.72	-0.87	8.89	22.67	9.28
101	-9.3	13.0	-15.3	0.68	1.29	-0.67	12.86	24.99	11.13
102	-8.5	15.2	-15.5	1.35	2.15	-0.43	16.86	26.10	13.31
103	-7.3	17.9	-15.4	1.68	2.54	-0.17	19.85	24.82	15.13
104	-5.9	20.6	-15.0	1.92	2.82	-0.09	20.33	20.07	15.75
105	-2.2	23.4	-14.8	2.01	2.92	0.32	17.25	11.29	14.73
106	0.3	26.0	-13.8	1.89	2.79	0.50	9.96	-0.98	12.09
107	1.4	28.1	-13.2	1.63	2.44	0.60	-0.86	-14.53	8.17
108					1.96	0.62	-11.47	-24.75	3.56
109						0.58	-13.88	-23.83	-0.86

176

APPENDIX B
Kinetic Analyses

Table B.1 Joint Reaction Forces, Net Muscle Moment—Right Ankle

POS	TIME	MODE	RFX (NT)	RFY (NT) GROUND	RFX (NT) R. ANKLE	RFY (NT) R. ANKLE	MOM. (N-M) GROUND	MOM. (N-M) R. ANKLE	MOM. (N-M) INERTIAL	FX (NT) INERTIAL	FY (NT) INERTIAL	MODE
1	0.00	HSR	15.420	18.510	-15.226	-11.149	-0.000	-1.438	-0.413	0.194	7.361	HSR
2	0.02		-28.610	201.620	26.947	-190.946	-0.000	8.656	-0.303	-1.663	10.674	
3	0.03		-79.140	387.380	75.250	-373.520	-0.000	20.978	-0.125	-3.890	13.860	
4	0.05		-131.090	557.420	125.256	-541.027	-0.000	33.870	0.077	-5.834	16.393	
5	0.07		-172.150	689.040	165.138	-671.146	-0.000	36.144	0.243	-7.012	17.894	
6	0.08		-195.260	775.690	188.010	-757.406	-0.000	41.419	0.341	-7.250	18.283	
7	0.10		-202.190	829.940	195.526	-812.145	-0.000	40.640	0.367	-6.664	17.794	
8	0.12		-197.720	866.880	192.220	-850.119	-0.000	36.342	0.335	-5.500	16.761	
9	0.13		-185.820	890.290	181.784	-874.819	-0.000	27.714	0.268	-4.036	15.471	
10	0.15	TOL	-169.300	895.560	166.786	-881.384	-0.000	20.834	0.190	-2.514	14.176	TOL
11	0.17		-149.960	878.730	148.847	-865.668	-0.000	11.443	0.120	-1.113	13.061	
12	0.18		-129.130	840.910	129.151	-828.708	-0.000	0.829	0.066	0.021	12.202	
13	0.20		-108.330	788.510	109.103	-776.898	-0.000	-8.262	0.027	0.773	11.612	
14	0.22		-89.350	731.160	90.466	-719.861	-0.000	-18.619	0.001	1.116	11.299	
15	0.23		-73.730	677.980	74.855	-666.773	-0.000	-30.240	-0.012	1.125	11.207	
16	0.25		-61.760	634.740	62.678	-623.519	-0.000	-36.602	-0.018	0.918	11.221	
17	0.27		-52.250	603.900	52.898	-592.631	-0.000	-43.935	-0.020	0.648	11.268	
18	0.28		-43.870	585.630	44.303	-574.275	-0.000	-50.182	-0.022	0.433	11.355	
19	0.30		-36.060	578.600	36.313	-567.105	-0.000	-57.693	-0.022	0.253	11.495	
20	0.32		-28.590	580.960	28.604	-569.308	-0.000	-61.811	-0.022	0.014	11.652	
21	0.33		-21.150	591.010	20.878	-579.242	-0.000	-67.541	-0.024	-0.272	11.768	
22	0.35		-13.370	607.130	12.909	-595.298	-0.000	-72.283	-0.030	-0.461	11.832	
23	0.37		-4.890	627.660	4.413	-615.782	-0.000	-77.700	-0.039	-0.477	11.878	
24	0.38		4.500	651.100	-4.900	-639.148	-0.000	-84.380	-0.051	-0.400	11.952	
25	0.40		14.980	676.180	-15.285	-664.092	-0.000	-88.073	-0.067	-0.305	12.088	
26	0.42		26.910	702.240	-27.014	-689.958	-0.000	-96.546	-0.085	-0.104	12.282	
27	0.43		40.580	729.170	-40.246	-716.686	-0.000	-103.098	-0.106	0.334	12.484	
28	0.45		55.860	756.840	-54.928	-744.205	-0.000	-113.094	-0.130	0.932	12.635	
29	0.47		72.390	784.090	-70.917	-771.379	-0.000	-118.527	-0.153	1.473	12.711	
30	0.48		89.600	808.280	-87.707	-795.545	-0.000	-125.178	-0.171	1.893	12.734	
31	0.50		106.640	825.890	-104.345	-813.161	-0.000	-133.009	-0.193	2.295	12.729	

Table B.1 (Continued)

POS	TIME	MODE	RFX (NT) GROUND	RFY (NT) GROUND	RFX (NT) R. ANKLE	RFY (NT) ANKLE	MOM. (N-M) GROUND	MOM. (N-M) R. ANKLE	MOM. (N-M) INERTIAL	FX (NT) INERTIAL	FY (NT) INERTIAL	MODE
32	0.52		122.730	833.530	-119.986	-820.818	-0.000	-135.583	-0.233	2.744	12.712	
33	0.53		137.450	828.180	-134.304	-815.436	-0.000	-137.393	-0.300	3.146	12.743	
34	0.55		150.320	806.790	-146.891	-793.846	-0.000	-136.317	-0.383	3.429	12.943	HSL
35	0.57		160.250	765.960	-156.480	-752.530	-0.000	-128.958	-0.452	3.770	13.429	
36	0.58	HSL	165.140	702.530	-160.531	-688.310	-0.000	-119.187	-0.473	4.609	14.220	
37	0.60		161.940	615.120	-155.536	-599.918	-0.000	-103.440	-0.419	6.404	15.202	
38	0.62		147.830	506.430	-138.504	-490.298	-0.000	-82.931	-0.273	9.326	16.132	
39	0.63		122.430	385.290	-109.283	-368.702	-0.000	-62.125	-0.037	13.147	16.588	
40	0.65		89.410	265.590	-72.057	-249.561	-0.000	-39.965	0.265	17.353	16.029	TOR
41	0.67		55.500	161.540	-34.209	-147.459	-0.000	-19.871	0.580	21.291	14.081	
42	0.68		27.220	82.110	-2.922	-71.250	-0.000	-6.064	0.839	24.298	10.860	
43	0.70		7.340	27.240	18.540	-20.251	-0.000	1.260	0.995	25.880	6.989	
44	0.72		-6.380	-12.010	32.191	15.281	-0.000	5.308	1.036	25.811	3.271	
45	0.73	TOR	-0.000	-0.000	24.190	0.348	-0.000	3.504	0.976	24.190	0.348	
46	0.75		-0.000	-0.000	21.468	-1.407	-0.000	3.115	0.850	21.468	-1.407	
47	0.77		-0.000	-0.000	18.315	-1.854	-0.000	2.615	0.695	18.315	-1.854	
48	0.78		-0.000	-0.000	15.318	-1.038	-0.000	2.111	0.532	15.318	-1.038	
49	0.80		-0.000	-0.000	12.740	0.889	-0.000	1.671	0.369	12.740	0.889	
50	0.82		-0.000	-0.000	10.509	3.695	-0.000	1.327	0.216	10.509	3.695	
51	0.83		-0.000	-0.000	8.413	7.012	-0.000	1.090	0.083	8.413	7.012	
52	0.85		-0.000	-0.000	6.331	10.363	-0.000	0.954	-0.021	6.331	10.363	
53	0.87		-0.000	-0.000	4.303	13.371	-0.000	0.902	-0.100	4.303	13.371	
54	0.88		-0.000	-0.000	2.368	15.921	-0.000	0.913	-0.157	2.368	15.921	
55	0.90		-0.000	-0.000	0.421	18.072	-0.000	0.966	-0.193	0.421	18.072	
56	0.92		-0.000	-0.000	-1.738	19.859	-0.000	1.037	-0.205	-1.738	19.859	
57	0.93		-0.000	-0.000	-4.280	21.232	-0.000	1.107	-0.193	-4.280	21.232	
58	0.95		-0.000	-0.000	-7.223	22.074	-0.000	1.164	-0.156	-7.223	22.074	
59	0.97		-0.000	-0.000	-10.428	22.189	-0.000	1.139	-0.103	-10.428	22.189	
60	0.98		-0.000	-0.000	-13.738	21.269	-0.000	0.954	-0.057	-13.738	21.269	
61	1.00		-0.000	-0.000	-17.050	19.023	-0.000	0.608	-0.050	-17.050	19.023	
62	1.02		-0.000	-0.000	-20.241	15.440	-0.000	0.134	-0.104	-20.241	15.440	
63	1.03		-0.000	-0.000	-23.076	10.913	-0.000	-0.380	-0.217	-23.076	10.913	
64	1.05		-0.000	-0.000	-25.187	6.134	-0.000	-0.825	-0.363	-25.187	6.134	
65	1.07		-0.000	-0.000	-26.171	1.955	-0.000	-1.096	-0.501	-26.171	1.955	
66	1.08		-0.000	-0.000	-25.758	-0.766	-0.000	-0.924	-0.598	-25.758	-0.766	
67	1.10		-0.000	-0.000	-23.965	-1.455	-0.000	-0.539	-0.631	-23.965	-1.455	
68	1.12	HSR	-0.000	-0.000	-21.056	-0.104	-0.000	-0.051	-0.601	-21.056	-0.104	
69	1.13		-0.000	-0.000	-17.423	2.769	-0.000		-0.520	-17.423	2.769	HSR
70	1.15		-0.000	-0.000	-13.608	6.440	-0.000		-0.400	-13.608	6.440	

179

Table B.1 (Continued)

POS	TIME	MODE	RFX (NT) GROUND	RFY (NT) GROUND	RFX (NT) R. ANKLE	RFY (NT) R. ANKLE	MOM. (N-M) GROUND	MOM. (N-M) R. ANKLE	MOM. (N-M) INERTIAL	FX (NT) INERTIAL	FY (NT) INERTIAL	MODE
71	1.17		-0.000	-0.000	-10.244	10.216	-0.000	0.471	-0.244	-10.244	10.216	
72	1.18		-0.000	-0.000	-7.701	13.507	-0.000	0.957	-0.063	-7.701	13.507	
73	1.20		-0.000	-0.000	-5.867	15.902	-0.000	1.345	0.115	-5.867	15.902	
74	1.22		-0.000	-0.000	-4.467	17.259	-0.000	1.597	0.253	-4.467	17.259	
75	1.23		-0.000	-0.000	-3.405	17.642	-0.000	1.705	0.332	-3.405	17.642	
76	1.25		-0.000	-0.000	-2.635	17.192	-0.000	1.682	0.349	-2.635	17.192	
77	1.27		-0.000	-0.000	-2.002	16.140	-0.000	1.563	0.312	-2.002	16.140	
78	1.28		-0.000	-0.000	-1.387	14.813	-0.000	1.392	0.237	-1.387	14.813	
79	1.30		-0.000	-0.000	-0.842	13.514	-0.000	1.211	0.145	-0.842	13.514	
80	1.32	TOL	-0.000	-0.000	-0.467	12.468	-0.000	1.050	0.055	-0.467	12.468	TOL
81	1.33		-0.000	-0.000	-0.273	11.797	-0.000	0.931	-0.015	-0.273	11.797	
82	1.35		-0.000	-0.000	-0.239	11.459	-0.000	0.865	-0.053	-0.239	11.459	
83	1.37		-0.000	-0.000	-0.342	11.300	-0.000	0.844	-0.053	-0.342	11.300	
84	1.38		-0.000	-0.000	-0.454	11.233	-0.000	0.854	-0.029	-0.454	11.233	
85	1.40		-0.000	-0.000	-0.404	11.275	-0.000	0.880	-0.005	-0.404	11.275	
86	1.42		-0.000	-0.000	-0.183	11.440	-0.000	0.910	0.004	-0.183	11.440	
87	1.43		-0.000	-0.000	0.032	11.658	-0.000	0.932	0.001	0.032	11.658	
88	1.45		-0.000	-0.000	0.078	11.822	-0.000	0.938	-0.004	0.078	11.822	
89	1.47		-0.000	-0.000	-0.021	11.892	-0.000	0.928	-0.011	-0.021	11.892	
90	1.48		-0.000	-0.000	-0.136	11.902	-0.000	0.910	-0.018	-0.136	11.902	
91	1.50		-0.000	-0.000	-0.201	11.908	-0.000	0.895	-0.026	-0.201	11.908	
92	1.52		-0.000	-0.000	-0.218	11.927	-0.000	0.885	-0.032	-0.218	11.927	
93	1.53		-0.000	-0.000	-0.171	11.960	-0.000	0.879	-0.039	-0.171	11.960	
94	1.55		-0.000	-0.000	0.241	12.011	-0.000	0.872	-0.053	0.241	12.011	
95	1.57		-0.000	-0.000	0.067	12.067	-0.000	0.862	-0.076	0.067	12.067	
96	1.58		-0.000	-0.000	0.617	12.126	-0.000	0.852	-0.107	0.617	12.126	
97	1.60		-0.000	-0.000	1.096	12.229	-0.000	0.846	-0.140	1.096	12.229	
98	1.62		-0.000	-0.000	1.638	12.384	-0.000	0.849	-0.172	1.638	12.384	
99	1.63		-0.000	-0.000	2.211	12.528	-0.000	0.855	-0.198	2.211	12.528	
100	1.65		-0.000	-0.000	2.770	12.628	-0.000	0.853	-0.225	2.770	12.628	
101	1.67		-0.000	-0.000	3.225	12.735	-0.000	0.838	-0.255	3.225	12.735	
102	1.68		-0.000	-0.000	3.437	12.874	-0.000	0.806	-0.280	3.437	12.874	
103	1.70		-0.000	-0.000	3.238	12.902	-0.000	0.749	-0.284	3.238	12.902	
104	1.72	HSL	-0.000	-0.000	2.507	12.565	-0.000	0.660	-0.248	2.507	12.565	HSL
105	1.73		-0.000	-0.000	1.243	11.743	-0.000	0.555	-0.150	1.243	11.743	
106	1.75		-0.000	-0.000	-0.366	10.645	-0.000	0.475	0.019	-0.366	10.645	
107	1.77		-0.000	-0.000	-1.900	9.708	-0.000	0.443	0.222	-1.900	9.708	
108	1.78		-0.000	-0.000	-2.679	9.390	-0.000	0.451	0.371	-2.679	9.390	
109	1.80		-0.000	-0.000	-1.591	10.242	-0.000	0.497	0.342	-1.591	10.242	

Table B.2 Joint Reaction Forces, Net Muscle Moment—Right Knee

POS	TIME	MODE	RFX (NT) R. ANKLE	RFY (NT) R. ANKLE	RFX (NT) R. KNEE	RFY (NT) R. KNEE	MOM. (N-M) R. ANKLE	MOM. (N-M) R. KNEE	MOM. (N-M) INERTIAL	FX (NT) INERTIAL	FY (NT) INERTIAL	MODE
1	0.00	HSR	15.226	11.149	10.381	29.105	1.438	-3.620	-1.373	25.607	40.254	HSR
2	0.02		-26.947	190.946	41.873	-146.808	-8.656	-0.400	-1.061	14.926	44.138	
3	0.03		-75.250	373.520	77.148	-327.528	-20.978	13.208	-0.750	1.898	45.992	
4	0.05		-125.256	541.027	114.148	-495.571	-33.870	36.835	-0.415	-11.108	45.456	
5	0.07		-165.138	671.146	142.825	-628.156	-36.144	55.760	-0.031	-22.313	42.990	
6	0.08		-188.010	757.406	157.654	-717.728	-41.419	78.684	0.366	-30.356	39.679	
7	0.10		-195.526	812.145	161.090	-775.311	-40.640	93.559	0.700	-34.435	36.835	
8	0.12		-192.220	850.119	157.772	-814.826	-36.342	102.017	0.918	-34.448	35.292	
9	0.13		-181.784	874.819	150.717	-839.816	-27.714	103.218	1.009	-31.067	35.002	
10	0.15	TOL	-166.786	881.384	141.277	-845.973	-20.834	103.143	0.997	-25.508	35.411	TOL
11	0.17		-148.847	865.668	129.695	-829.625	-11.443	97.160	0.919	-19.152	36.044	
12	0.18		-129.151	828.708	115.931	-792.074	-0.829	86.376	0.812	-13.220	36.634	
13	0.20		-109.103	776.898	100.603	-739.873	8.262	74.055	0.694	-8.500	37.025	
14	0.22		-90.466	719.861	85.280	-682.745	18.619	58.703	0.568	-5.186	37.116	
15	0.23		-74.855	666.773	71.857	-629.913	30.240	41.900	0.436	-2.998	36.860	
16	0.25		-62.678	623.519	61.201	-589.137	36.602	31.210	0.298	-1.477	36.382	
17	0.27		-52.898	592.631	52.742	-556.585	43.935	20.631	0.163	-0.156	36.047	
18	0.28		-44.303	574.275	45.460	-538.044	50.182	12.129	0.048	1.158	36.230	
19	0.30		-36.313	567.105	38.516	-530.105	57.693	3.269	-0.042	2.204	36.999	
20	0.32		-28.604	569.308	31.159	-531.328	61.811	-1.395	-0.109	2.555	37.980	
21	0.33		-20.878	579.242	23.085	-540.603	67.541	-6.902	-0.154	2.207	38.639	
22	0.35		-12.909	595.298	14.669	-556.537	72.283	-10.608	-0.182	1.761	38.761	
23	0.37		-4.413	615.782	6.278	-577.194	77.700	-14.238	-0.207	1.866	38.588	
24	0.38		-4.900	639.148	-2.240	-600.604	84.380	-18.573	-0.250	2.660	38.544	
25	0.40		15.285	664.092	-11.318	-625.229	88.073	-19.552	-0.315	3.968	38.863	
26	0.42		27.014	689.958	-21.178	-650.517	96.546	-24.957	-0.388	5.837	39.442	
27	0.43		40.246	716.686	-31.778	-676.692	103.098	-27.915	-0.447	8.468	39.993	
28	0.45		54.928	744.205	-43.284	-703.933	113.094	-33.637	-0.491	11.644	40.272	
29	0.47		70.917	771.379	-56.036	-731.249	118.527	-34.162	-0.532	14.881	40.130	
30	0.48		87.707	795.545	-69.665	-756.086	125.178	-35.382	-0.583	18.042	39.459	
31	0.50		104.345	813.161	-82.954	-774.894	133.009	-37.341	-0.645	21.391	38.267	

Table B.2 (Continued)

POS	TIME	MODE	RPX (NT) R. ANKLE	RPY (NT) R. ANKLE	RPX (NT) R. KNEE	RPY (NT) R. KNEE	MOM. (N-M) R. ANKLE	MOM. (N-M) R. KNEE	MOM. (N-M) INERTIAL	FX (NT) INERTIAL	FY (NT) INERTIAL	MODE
32	0.52		119.986	820.818	-94.889	-783.952	135.583	-33.920	-0.710	25.097	36.866	
33	0.53		134.304	815.436	-105.425	-779.687	137.393	-30.431	-0.773	28.880	35.749	
34	0.55		146.891	793.846	-114.612	-758.598	136.317	-25.853	-0.833	32.279	35.248	HSL
35	0.57		156.480	752.530	-121.355	-717.063	128.958	-17.876	-0.866	35.124	35.467	
36	0.58	HSL	160.531	688.310	-122.875	-651.914	119.187	-11.218	-0.824	37.657	36.396	
37	0.60		155.536	599.918	-115.446	-562.031	103.440	-2.870	-0.668	40.090	37.887	
38	0.62		138.504	490.298	-96.277	-450.634	82.931	5.793	-0.400	42.227	39.664	
39	0.63		109.283	368.702	-66.039	-327.342	62.125	10.931	-0.040	43.244	41.360	
40	0.65		72.057	249.561	-30.123	-207.109	39.965	15.287	0.419	41.934	42.452	
41	0.67		34.209	147.459	3.384	-105.045	19.871	17.864	0.994	37.593	42.414	
42	0.68		2.922	71.250	27.852	-30.179	6.064	16.691	1.657	30.774	41.072	
43	0.70		-18.540	20.251	41.672	18.380	-1.260	12.495	2.316	23.132	38.632	
44	0.72		-32.191	-15.281	48.444	50.640	-5.308	7.358	2.864	16.253	35.360	
45	0.73	TOR	-24.190	-0.348	34.852	31.815	-3.504	8.946	3.230	10.662	31.467	TOR
46	0.75		-21.468	1.407	27.630	25.955	-3.115	8.791	3.293	6.662	27.362	
47	0.77		-18.315	1.854	21.026	21.844	-2.615	7.938	3.357	2.711	23.698	
48	0.78		-15.318	1.038	15.924	20.077	-2.111	6.537	3.146	0.606	21.115	
49	0.80		-12.740	-0.889	12.619	20.971	-1.671	4.745	2.799	-0.121	20.082	
50	0.82		-10.509	-3.695	10.388	24.486	-1.327	2.662	2.369	-0.121	20.791	
51	0.83		-8.413	-7.012	8.017	29.952	-1.090	0.415	1.913	-0.395	22.940	
52	0.85		-6.331	-10.363	4.772	36.142	-0.954	-1.758	1.484	-1.560	25.779	
53	0.87		-4.303	-13.371	0.772	42.023	-0.902	-3.586	1.126	-3.425	28.652	
54	0.88		-2.368	-15.921	-3.121	47.425	-0.913	-4.958	0.843	-5.489	31.504	
55	0.90		-0.421	-18.072	-7.183	52.705	-0.966	-5.983	0.600	-7.603	34.633	
56	0.92		1.738	-19.859	-11.702	57.810	-1.037	-6.837	0.354	-9.964	37.951	
57	0.93		4.280	-21.702	-17.035	62.081	-1.107	-7.652	0.077	-12.754	40.850	
58	0.95		7.223	-22.074	-23.100	64.886	-1.164	-8.482	-0.246	-15.877	42.812	
59	0.97		10.428	-22.189	-29.363	65.934	-1.189	-9.331	-0.635	-18.935	43.745	
60	0.98		13.738	-21.269	-35.177	64.983	-1.139	-10.242	-1.112	-21.439	43.714	
61	1.00		17.050	-19.023	-40.106	61.780	-0.954	-11.353	-1.682	-23.056	42.756	
62	1.02		20.241	-15.440	-44.059	56.439	-0.608	-12.800	-2.296	-23.818	40.998	
63	1.03		23.076	-10.913	-47.181	49.628	-0.134	-14.543	-2.874	-24.032	38.715	
64	1.05		25.187	-6.134	-49.181	42.329	0.380	-16.286	-3.327	-23.993	36.245	
65	1.07		26.171	-1.955	-50.021	35.880	0.825	-17.547	-3.581	-23.850	33.925	
66	1.08		25.758	0.766	-49.366	31.425	1.129	-17.823	-3.580	-23.608	32.191	
67	1.10		23.965	1.455	-47.123	30.169	0.924	-16.801	-3.307	-23.158	31.623	
68	1.12	HSR	21.056	0.104	-43.315	32.531	-0.924	-14.514	-2.803	-22.258	32.634	
69	1.13		17.423	-2.769	-38.058	37.904	0.539	-11.324	-2.166	-20.635	35.135	
70	1.15		13.608	-6.440	-32.109	44.964	0.051	-7.869	-1.518	-18.501	38.524	HSR

Table B.2 (Continued)

POS	TIME	MODE	RPX (NT) R. ANKLE	RPY (NT) ANKLE	RPX (NT) R. KNEE	RPY (NT) R. KNEE	MOM. (N-M) R. ANKLE	MOM. (N-M) R. KNEE	MOM. (N-M) INERTIAL	FX (NT) INERTIAL	FY (NT) INERTIAL	MODE
71	1.17		10.244	-10.216	-27.055	51.976	-0.471	-4.927	-0.957	-16.811	41.759	
72	1.18		7.701	-13.507	-24.110	57.137	-0.957	-3.016	-0.511	-16.409	43.630	
73	1.20		5.867	-15.902	-23.140	59.388	-1.345	-2.101	-0.139	-17.273	43.486	
74	1.22		4.467	-17.259	-23.387	59.006	-1.597	-1.869	0.197	-18.919	41.747	
75	1.23		3.405	-17.642	-24.241	57.098	-1.705	-2.074	0.483	-20.836	39.456	
76	1.25		2.635	-17.192	-24.836	54.690	-1.682	-2.470	0.679	-22.201	37.498	
77	1.27		2.002	-16.140	-23.973	52.474	-1.563	-2.721	0.779	-21.971	36.334	
78	1.28		1.387	-16.813	-21.244	50.844	-1.392	-2.640	0.806	-19.857	36.031	
79	1.30		0.842	-13.514	-17.561	49.822	-1.211	-2.361	0.791	-16.719	36.307	
80	1.32	TOL	0.467	-12.468	-14.194	49.313	-1.050	-2.125	0.755	-13.727	36.846	TOL
81	1.33		0.273	-11.797	-11.745	49.290	-0.931	-2.043	0.705	-11.472	37.493	
82	1.35		0.239	-11.459	-10.170	49.477	-0.865	-2.104	0.638	-9.932	38.018	
83	1.37		0.342	-11.300	-9.145	49.442	-0.844	-2.246	0.547	-8.803	38.142	
84	1.38		0.454	-11.233	-8.117	49.147	-0.854	-2.335	0.444	-7.664	37.915	
85	1.40		0.404	-11.275	-6.578	48.950	-0.880	-2.242	0.356	-6.175	37.676	
86	1.42		0.183	-11.440	-4.508	49.063	-0.910	-1.987	0.283	-4.325	37.622	
87	1.43		-0.032	-11.658	-2.453	49.303	-0.932	-1.747	0.204	-2.484	37.645	
88	1.45		-0.078	-11.822	-0.953	49.435	-0.938	-1.674	0.105	-1.030	37.613	
89	1.47		-0.021	-11.892	0.056	49.525	-0.928	-1.747	-0.006	0.077	37.633	
90	1.48		0.136	-11.902	1.108	49.777	-0.910	-1.832	-0.123	1.244	37.875	
91	1.50		0.201	-11.908	2.542	50.218	-0.895	-1.846	-0.239	2.743	38.310	
92	1.52		0.218	-11.927	4.126	50.684	-0.885	-1.816	-0.339	4.344	38.757	
93	1.53		0.171	-11.960	5.557	51.026	-0.879	-1.780	-0.405	5.728	39.066	
94	1.55		0.021	-12.011	6.980	51.137	-0.872	-1.701	-0.435	7.001	39.126	
95	1.57		-0.241	-12.067	8.858	50.945	-0.862	-1.513	-0.453	8.618	38.878	
96	1.58		-0.617	-12.126	11.664	50.638	-0.852	-1.169	-0.492	11.048	38.512	
97	1.60		-1.096	-12.229	15.493	50.508	-0.846	-0.684	-0.562	14.397	38.279	
98	1.62		-1.638	-12.384	19.894	50.335	-0.849	-0.122	-0.640	18.256	37.951	
99	1.63		-2.211	-12.528	24.281	49.650	-0.855	0.449	-0.702	22.070	37.122	
100	1.65		-2.770	-12.628	28.176	48.639	-0.853	0.923	-0.736	25.406	36.011	
101	1.67		-3.225	-12.735	30.896	48.046	-0.838	1.098	-0.730	27.671	35.311	
102	1.68		-3.437	-12.874	31.334	48.261	-0.806	0.730	-0.666	27.897	35.387	
103	1.70		-3.238	-12.902	28.198	48.841	-0.749	-0.357	-0.521	24.960	35.938	
104	1.72	HSL	-2.507	-12.565	20.368	48.859	-0.660	-2.174	-0.263	17.860	36.294	
105	1.73		-1.243	-11.743	7.290	47.675	-0.555	-4.575	0.134	6.047	35.932	HSL
106	1.75		-0.366	-10.645	-10.157	45.549	-0.475	-7.281	0.648	-9.791	34.905	
107	1.77		1.900	-9.708	-28.435	43.543	-0.443	-9.839	1.172	-26.535	33.835	
108	1.78		2.679	-9.390	-40.250	42.872	-0.451	-11.469	1.500	-37.570	33.483	
109	1.80		1.591	-10.242	-33.450	44.812	-0.497	-10.862	1.331	-31.858	34.570	

Table B.3 Joint Reaction Forces, Net Muscle Moment—Right Hip

POS	TIME	MODE	BFX (NT) R. KNEE	RFY (NT) R. KNEE	RFX (NT) R. HIP	RFY (NT) R. HIP	MOM. (N-M) R. KNEE	MOM. (N-M) R. HIP	MOM. (N-M) INERTIAL	PX (NT) INERTIAL	PY (NT) INERTIAL	MODE
1	0.00	HSR	-10.381	-29.105	99.299	125.008	3.620	25.781	-0.226	88.918	95.904	HSR
2	0.02		-41.873	146.808	102.176	-45.993	0.400	6.931	-0.304	60.303	100.815	
3	0.03		-77.148	327.528	103.884	-225.623	-13.208	-1.940	-0.398	26.736	101.905	
4	0.05		-114.148	495.571	108.153	-397.088	-36.835	1.845	-0.603	-5.995	98.483	
5	0.07		-142.825	628.156	109.032	-536.628	-55.760	4.509	-0.984	-33.793	91.528	
6	0.08		-157.654	717.728	104.169	-634.087	-78.684	14.987	-1.500	-53.485	83.640	
7	0.10		-161.090	775.311	97.626	-697.500	-93.559	21.300	-2.007	-63.465	77.810	
8	0.12		-157.772	814.826	93.368	-739.307	-102.017	25.073	-2.335	-64.404	75.519	
9	0.13		-150.717	839.816	91.896	-763.933	-103.218	25.903	-2.369	-58.821	75.883	
10	0.15	TOL	-141.277	845.973	91.407	-769.048	-103.143	30.167	-2.093	-49.870	76.925	TOL
11	0.17		-129.695	829.625	89.372	-752.397	-97.160	32.561	-1.592	-40.323	77.227	
12	0.18		-115.931	792.074	83.921	-715.622	-86.376	32.450	-1.007	-32.010	76.452	
13	0.20		-100.603	739.873	74.876	-665.012	-74.055	31.123	-0.465	-25.727	74.861	
14	0.22		-85.280	682.745	64.045	-609.893	-58.703	25.749	-0.042	-21.235	72.851	
15	0.23		-71.857	629.913	54.298	-559.271	-41.900	17.555	0.236	-17.559	70.643	
16	0.25		-61.201	587.137	47.518	-518.675	-31.210	14.286	0.388	-13.683	68.462	
17	0.27		-52.742	556.585	43.579	-489.611	-20.631	10.085	0.460	-9.163	66.973	
18	0.28		-45.460	538.044	41.160	-470.941	-12.129	6.968	0.481	-4.300	67.104	
19	0.30		-38.516	530.105	38.640	-460.965	-3.269	2.506	0.458	0.123	69.141	
20	0.32		-31.159	531.328	34.368	-459.184	1.395	1.172	0.386	3.210	72.145	
21	0.33		-23.085	540.603	27.928	-465.858	6.902	-1.887	0.264	4.842	74.744	
22	0.35		-14.669	556.537	20.732	-480.311	10.608	-3.331	0.120	6.063	76.226	
23	0.37		-6.278	577.194	14.324	-500.433	14.238	-4.192	0.008	8.045	76.761	
24	0.38		2.240	600.604	8.762	-523.596	18.573	-5.060	-0.029	11.002	77.008	
25	0.40		11.318	625.229	3.079	-547.768	19.552	-2.062	0.028	14.397	77.460	
26	0.42		21.178	650.517	-3.304	-572.459	24.957	-3.141	0.181	17.874	78.057	
27	0.43		31.778	676.692	-10.440	-598.354	27.915	-1.467	0.422	21.338	78.338	
28	0.45		43.284	703.933	-18.640	-625.958	33.637	2.460	0.722	24.644	77.975	
29	0.47		56.036	731.249	-28.223	-654.158	34.162	1.414	1.037	27.812	77.090	
30	0.48		69.665	756.086	-38.259	-680.306	35.382	4.020	1.333	31.406	75.780	
31	0.50		82.954	774.894	-46.952	-701.206	37.341	5.264	1.605	36.002	73.688	

Table B.3 (Continued)

POS	TIME	MODE	BFX (NT) R. KNEE	RFY (NT)	RFX (NT) R. HIP	RFY (NT)	MOM. (N-M) R. KNEE	MOM. (N-M) R. HIP	MOM. (N-M) INERTIAL	FX (NT) INERTIAL	FY (NT) INERTIAL	MODE
32	0.52		94.889	783.952	-53.407	-713.199	33.920	10.916	1.859	41.482	70.752	
33	0.53		105.425	779.687	-58.421	-712.007	30.431	14.754	2.089	47.003	67.680	
34	0.55		114.612	758.598	-63.253	-691.374	25.853	16.868	2.304	51.359	65.224	
35	0.57		121.355	717.063	-67.781	-652.984	17.876	19.268	2.523	53.574	64.079	
36	0.58	HSL	122.875	651.914	-69.458	-586.628	11.218	17.957	2.753	53.416	65.286	HSL
37	0.60		115.446	562.031	-64.204	-492.650	2.870	17.668	2.970	51.242	69.381	
38	0.62		96.277	450.634	-48.993	-374.936	-5.793	19.276	3.143	47.285	75.698	
39	0.63		66.039	327.342	-24.893	-244.051	-10.931	20.673	3.222	41.146	83.290	
40	0.65		30.123	207.109	1.866	-115.616	-15.287	24.761	3.113	31.989	91.493	
41	0.67		-3.384	105.045	23.110	-5.342	-17.864	29.164	2.730	19.726	99.703	
42	0.68		-27.852	30.179	33.970	30.091	-16.691	30.091	2.077	6.118	107.365	
43	0.70		-41.672	-18.380	35.787	132.329	-12.495	27.373	1.278	-5.885	113.949	
44	0.72	TOR	-48.444	-50.640	34.143	168.924	-7.358	23.645	0.482	-14.301	118.284	TOR
45	0.73		-34.852	-31.815	15.421	150.579	-8.946	20.001	-0.238	-19.431	118.764	
46	0.75		-27.630	-25.955	4.934	140.514	-8.791	17.130	-0.867	-22.696	114.560	
47	0.77		-21.026	-21.844	-3.899	128.111	-7.938	13.783	-1.360	-24.925	106.267	
48	0.78		-15.924	-20.077	-10.222	115.480	-6.537	10.589	-1.660	-26.146	95.403	
49	0.80		-12.619	-20.971	-13.630	104.858	-4.745	7.999	-1.768	-26.249	83.887	
50	0.82		-10.388	-24.486	-15.381	97.962	-2.662	5.861	-1.745	-25.769	73.475	
51	0.83		-8.017	-29.952	-17.758	94.895	-0.415	3.664	-1.673	-25.775	64.943	
52	0.85		-4.772	-36.142	-21.950	94.192	1.758	1.155	-1.634	-26.722	58.050	
53	0.87		-0.878	-42.023	-27.024	94.661	3.586	-1.319	-1.663	-27.902	52.638	
54	0.88		3.121	-47.425	-31.489	96.682	4.958	-3.236	-1.716	-28.368	49.257	
55	0.90		7.183	-52.705	-34.988	101.056	5.983	-4.495	-1.716	-27.806	48.352	
56	0.92		11.702	-57.810	-37.800	107.136	6.837	-5.428	-1.635	-26.098	49.326	
57	0.93		17.035	-62.081	-40.176	112.957	7.652	-6.505	-1.500	-23.142	50.876	
58	0.95		23.100	-64.886	-42.229	117.072	8.482	-7.972	-1.323	-19.129	52.186	
59	0.97		29.363	-65.934	-43.355	119.615	9.331	-9.642	-1.092	-13.992	53.681	
60	0.98		35.177	-64.983	-42.132	121.675	10.242	-11.163	-0.827	-6.955	56.692	
61	1.00		40.106	-61.780	-37.712	123.986	11.353	-12.445	-0.576	2.074	62.206	
62	1.02		44.059	-56.439	-31.185	126.203	12.800	-13.789	-0.356	12.874	69.765	
63	1.03		47.108	-49.628	-24.954	127.130	14.543	-15.538	-0.140	22.154	77.501	
64	1.05		49.181	-42.379	-21.032	125.868	16.286	-17.639	0.090	28.149	83.489	
65	1.07		50.021	-35.880	-20.068	122.751	17.547	-19.554	0.290	29.952	86.871	
66	1.08		49.366	-31.169	-21.622	119.654	16.801	-20.480	0.372	27.744	88.229	
67	1.10		47.123	-30.169	-24.791	119.261	16.514	-19.719	0.266	22.332	89.093	
68	1.12	HSR	43.315	-32.531	-28.259	123.386	14.514	-16.919	-0.002	15.055	90.856	
69	1.13		38.058	-37.904	-30.568	131.860	11.334	-12.209	-0.322	7.490	93.956	HSR
70	1.15		32.109	-44.964	-31.677	142.596	7.869	-6.544	-0.574	0.432	97.632	

Table B.3 (Continued)

POS	TIME	MODE	BFX (NT) R. KNEE	RFY (NT) R. KNEE	RFX (NT) R. HIP	RFY (NT) R. HIP	MOM. (N-M) R. KNEE	MOM. (N-M) R. HIP	MOM. (N-M) INERTIAL	FX (NT) INERTIAL	FY (NT) INERTIAL	MODE
71	1.17		27.055	-51.976	-33.619	152.043	4.927	-1.737	-0.705	-6.564	100.067	
72	1.18		24.110	-57.137	-38.239	156.313	3.016	0.651	-0.763	-14.129	99.175	
73	1.20		23.140	-59.388	-45.246	153.756	2.101	0.402	-0.864	-22.106	94.367	
74	1.22		23.367	-59.006	-53.133	146.597	1.869	-1.670	-1.089	-29.747	87.591	
75	1.23		24.241	-57.098	-60.586	138.612	2.074	-4.593	-1.395	-36.345	81.514	
76	1.25		24.836	-54.690	-65.940	132.068	2.470	-7.285	-1.623	-41.105	77.378	
77	1.27		23.973	-52.474	-66.944	127.602	2.721	-8.549	-1.633	-42.970	75.128	
78	1.28		21.244	-50.844	-62.905	125.451	2.640	-7.944	-1.435	-41.660	74.607	
79	1.30		17.561	-49.822	-55.819	125.314	2.361	-6.248	-1.155	-38.258	75.492	
80	1.32	TOL	14.194	-49.313	-48.461	126.342	2.125	-4.605	-0.910	-34.267	77.028	TOL
81	1.33		11.745	-49.290	-42.301	127.683	2.043	-3.591	-0.740	-30.556	78.393	
82	1.35		10.170	-49.477	-37.386	128.501	2.104	-3.230	-0.615	-27.216	79.024	
83	1.37		9.145	-49.442	-33.157	127.938	2.246	-3.288	-0.485	-24.013	78.496	
84	1.38		8.117	-49.147	-28.865	125.956	2.335	-3.334	-0.331	-20.748	76.809	
85	1.40		6.578	-48.950	-23.760	123.571	2.242	-2.945	-0.189	-17.181	74.621	
86	1.42		4.508	-49.063	-17.491	121.852	1.987	-2.070	-0.102	-12.984	72.789	
87	1.43		2.453	-49.303	-10.663	121.118	1.747	-1.132	-0.076	-8.210	71.816	
88	1.45		0.953	-49.435	-4.361	121.244	1.674	-0.603	-0.086	-3.210	71.809	
89	1.47		-0.056	-49.525	1.057	122.088	1.747	-0.479	-0.082	1.001	72.563	
90	1.48		-1.108	-49.777	6.094	123.575	1.832	-0.369	-0.006	4.986	73.798	
91	1.50		-2.542	-50.218	10.944	125.525	1.846	-0.053	0.175	8.402	75.307	
92	1.52		-4.126	-50.684	14.922	127.644	1.816	0.234	0.426	10.796	76.960	
93	1.53		-5.557	-51.026	17.800	129.426	1.780	0.282	0.655	12.243	78.400	
94	1.55		-6.980	-51.137	20.773	130.154	1.701	0.343	0.791	13.793	79.017	
95	1.57		-8.858	-50.945	25.614	129.331	1.513	0.957	0.846	16.756	78.386	
96	1.58		-11.664	-50.638	33.599	127.433	1.572	2.572	0.892	21.934	76.795	
97	1.60		-15.493	-50.508	44.663	125.451	0.684	5.190	0.976	29.171	74.943	
98	1.62		-19.894	-50.335	57.041	123.412	0.122	8.342	1.097	37.147	73.078	
99	1.63		-24.281	-49.650	68.342	120.636	-0.449	11.483	1.264	44.061	70.986	
100	1.65		-28.176	-48.639	76.551	117.450	-0.923	14.075	1.495	48.375	68.811	
101	1.67		-30.896	-48.046	79.635	115.301	-1.098	15.315	1.759	48.739	67.254	
102	1.68		-31.334	-48.261	75.354	115.419	-0.730	14.180	1.967	44.020	67.158	
103	1.70		-28.198	-48.841	61.333	117.796	0.357	9.747	2.017	33.135	68.955	
104	1.72	HSL	-20.368	-48.859	34.737	121.093	2.174	1.353	1.863	14.369	72.234	
105	1.73		-7.290	-47.675	-6.620	123.921	4.575	-11.243	1.512	-13.910	76.246	
106	1.75		10.157	-45.549	-60.611	126.302	7.281	-27.222	1.010	-50.453	80.753	
107	1.77		28.435	-43.543	-116.502	129.220	9.839	-43.597	0.444	-88.067	85.677	
108	1.78		40.250	-42.872	-151.753	132.747	11.469	-54.079	-0.014	-111.504	89.875	HSL
109	1.80		33.450	-44.812	-129.679	135.043	10.862	-48.113	-0.026	-96.229	90.231	

Table B.4 Kinetic, Potential and Total Energy—Right Foot

INSTANTANEOUS ENERGY (JOULES) OF R. FOOT   WW2IF2   NORMAL WALKING SPEED   FILTERED AT 4 HZ   TRUNK & FORCES MOD.

POS	TRANSLATIONAL	ROTATIONAL	POTENTIAL	TOTAL	MODE1	MODE2	X KINETIC	Y KINETIC
1	0.43	0.03	1.34	1.80	HSR	1	0.23	0.19
2	0.44	0.05	1.23	1.72		1	0.23	0.21
3	0.41	0.06	1.11	1.56		1	0.20	0.20
4	0.32	0.06	1.00	1.39		1	0.15	0.17
5	0.22	0.05	0.91	1.18		1	0.10	0.12
6	0.13	0.04	0.83	1.00		1	0.06	0.07
7	0.06	0.02	0.78	0.86		1	0.03	0.04
8	0.02	0.01	0.74	0.77		1	0.01	0.02
9	0.01	0.01	0.71	0.73		1	0.00	0.01
10	0.00	0.00	0.70	0.70		1	0.00	0.00
11	0.00	0.00	0.70	0.70		1	0.00	0.00
12	0.00	0.00	0.70	0.70	TOL	1	0.00	0.00
13	0.00	0.00	0.70	0.70		1	0.00	0.00
14	0.00	0.00	0.70	0.70		1	0.00	0.00
15	0.00	0.00	0.70	0.70		1	0.00	0.00
16	0.00	0.00	0.70	0.70		1	0.00	0.00
17	0.00	0.00	0.70	0.70		1	0.00	0.00
18	0.00	0.00	0.70	0.70		1	0.00	0.00
19	0.00	0.00	0.70	0.70		1	0.00	0.00
20	0.00	0.00	0.70	0.71		1	0.00	0.00
21	0.00	0.00	0.71	0.71		1	0.00	0.00
22	0.00	0.00	0.71	0.71		1	0.00	0.00
23	0.00	0.00	0.72	0.72		1	0.00	0.00
24	0.00	0.00	0.72	0.73		1	0.00	0.00
25	0.00	0.00	0.73	0.74		1	0.00	0.00
26	0.00	0.00	0.74	0.75		1	0.00	0.00
27	0.00	0.01	0.76	0.77		1	0.00	0.00
28	0.00	0.01	0.77	0.79		1	0.00	0.01
29	0.01	0.01	0.79	0.82		1	0.00	0.01
30	0.01	0.01	0.80	0.82		1	0.00	0.01
31	0.02	0.02	0.82	0.86		1	0.01	0.01

Table B.4 (Continued)

INSTANTANEOUS ENERGY (JOULES) OF R. FOOT   WN21F2   NORMAL WALKING SPEED   FILTERED AT 4 HZ   TRUNK & FORCES MOD.

POS	TRANSLATIONAL	ROTATIONAL	POTENTIAL	TOTAL	MODE1	MODE2	X KINETIC	Y KINETIC
32	0.03	0.03	0.85	0.90		1	0.01	0.02
33	0.04	0.04	0.88	0.96		1	0.02	0.02
34	0.06	0.05	0.92	1.03		1	0.03	0.02
35	0.08	0.08	0.96	1.12		1	0.05	0.03
36	0.11	0.12	1.00	1.23		1	0.07	0.04
37	0.16	0.16	1.06	1.37	HSL	1	0.10	0.06
38	0.24	0.19	1.12	1.55		1	0.16	0.08
39	0.39	0.21	1.20	1.80		1	0.27	0.12
40	0.63	0.20	1.29	2.12		1	0.47	0.16
41	1.01	0.16	1.40	2.56		1	0.81	0.19
42	1.54	0.10	1.51	3.15		1	1.33	0.21
43	2.24	0.04	1.62	3.90		1	2.05	0.19
44	3.09	0.01	1.72	4.82	TOR	1	2.96	0.13
45	4.06	0.00	1.80	5.86		1	3.99	0.06
46	5.08	0.02	1.85	6.95		1	5.07	0.01
47	6.10	0.06	1.86	8.02		1	6.09	0.00
48	7.06	0.10	1.84	9.00		1	7.03	0.03
49	7.95	0.14	1.78	9.87		1	7.86	0.09
50	8.75	0.16	1.69	10.60		1	8.59	0.16
51	9.42	0.18	1.58	11.18		1	9.21	0.22
52	9.95	0.18	1.46	11.59		1	9.70	0.25
53	10.31	0.17	1.33	11.81		1	10.07	0.24
54	10.50	0.16	1.21	11.88		1	10.30	0.20
55	10.55	0.15	1.11	11.80		1	10.40	0.15
56	10.45	0.13	1.02	11.60		1	10.36	0.09
57	10.20	0.11	0.96	11.26		1	10.15	0.04
58	9.77	0.10	0.92	10.79		1	9.76	0.01
59	9.17	0.09	0.91	10.17		1	9.17	0.00
60	8.41	0.08	0.93	9.43		1	8.39	0.02
61	7.50	0.08	0.98	8.57		1	7.44	0.06
62	6.47	0.08	1.06	7.60		1	6.37	0.10
63	5.33	0.07	1.14	6.54		1	5.22	0.11
64	4.16	0.05	1.23	5.44		1	4.07	0.09
65	3.05	0.03	1.29	4.38		1	3.00	0.05
66	2.10	0.01	1.34	3.44		1	2.09	0.01
67	1.37	0.00	1.34	2.71		1	1.37	0.00
68	0.88	0.00	1.32	2.20		1	0.85	0.03
69	0.59	0.01	1.25	1.85	HSR	1	0.50	0.09
70	0.43	0.03	1.17	1.62		1	0.29	0.14

Table B.4 (Continued)

INSTANTANEOUS ENERGY (JOULES) OF R. FOOT   WN21F2   NORMAL WALKING SPEED   FILTERED AT 4 HZ   TRUNK & FORCES MOD.

POS	TRANSLATIONAL	ROTATIONAL	POTENTIAL	TOTAL	MODE1	MODE2	X KINETIC	Y KINETIC
71	0.33	0.04	1.07	1.44		1	0.17	0.16
72	0.25	0.05	0.97	1.27		1	0.10	0.16
73	0.18	0.05	0.87	1.11		1	0.06	0.13
74	0.12	0.04	0.79	0.95		1	0.03	0.09
75	0.07	0.03	0.73	0.82		1	0.02	0.05
76	0.04	0.01	0.68	0.73		1	0.01	0.02
77	0.02	0.01	0.65	0.67		1	0.01	0.01
78	0.01	0.00	0.63	0.64		1	0.00	0.01
79	0.00	0.00	0.62	0.63		1	0.00	0.00
80	0.00	0.00	0.62	0.62	TOL	1	0.00	0.00
81	0.00	0.00	0.62	0.62		1	0.00	0.00
82	0.00	0.00	0.62	0.62		1	0.00	0.00
83	0.00	0.00	0.62	0.62		1	0.00	0.00
84	0.00	0.00	0.62	0.62		1	0.00	0.00
85	0.00	0.00	0.62	0.62		1	0.00	0.00
86	0.00	0.00	0.62	0.62		1	0.00	0.00
87	0.00	0.00	0.62	0.62		1	0.00	0.00
88	0.00	0.00	0.62	0.62		1	0.00	0.00
89	0.00	0.00	0.62	0.62		1	0.00	0.00
90	0.00	0.00	0.63	0.63		1	0.00	0.00
91	0.00	0.00	0.63	0.63		1	0.00	0.00
92	0.00	0.00	0.64	0.64		1	0.00	0.00
93	0.00	0.00	0.65	0.65		1	0.00	0.00
94	0.30	0.00	0.66	0.66		1	0.00	0.00
95	0.00	0.00	0.67	0.67		1	0.00	0.00
96	0.00	0.00	0.68	0.69		1	0.00	0.00
97	0.01	0.01	0.70	0.71		1	0.00	0.00
98	0.01	0.01	0.71	0.73		1	0.00	0.01
99	0.01	0.01	0.73	0.76		1	0.00	0.01
100	0.02	0.02	0.76	0.80		1	0.01	0.01
101	0.03	0.03	0.79	0.85		1	0.02	0.01
102	0.05	0.04	0.82	0.90		1	0.03	0.02
103	0.06	0.06	0.85	0.97		1	0.04	0.02
104	0.08	0.07	0.89	1.05	HSL	1	0.06	0.03
105	0.10	0.09	0.93	1.12		1	0.07	0.03
106	0.10	0.09	0.98	1.17		1	0.07	0.03
107	0.09	0.09	1.02	1.19		1	0.06	0.03
108	0.07	0.06	1.06	1.19		1	0.05	0.02
109	0.05	0.04	1.09	1.18		1	0.04	0.01

Table B.5 Kinetic, Potential and Total Energy—Right Shank

INSTANTANEOUS ENERGY (JOULES) OF R. SHANK   WN21F2   NORMAL WALKING SPEED   FILTERED AT 4 HZ   TRUNK & FORCES MOD.

POS	TRANSLATIONAL	ROTATIONAL	POTENTIAL	TOTAL	MODE1	MODE2	X KINETIC	Y KINETIC
1	2.64	0.13	13.35	16.12	HSR	1	2.59	0.06
2	3.06	0.17	13.25	16.48		1	3.02	0.04
3	3.23	0.21	13.17	16.60		1	3.21	0.02
4	3.11	0.23	13.12	16.46		1	3.10	0.01
5	2.74	0.24	13.10	16.08		1	2.74	0.00
6	2.21	0.23	13.09	15.54		1	2.21	0.00
7	1.65	0.21	13.09	14.95		1	1.65	0.00
8	1.14	0.18	13.10	14.41		1	1.14	0.00
9	0.74	0.14	13.09	13.98		1	0.74	0.00
10	0.47	0.11	13.09	13.67	TOL	1	0.47	0.00
11	0.30	0.08	13.08	13.46		1	0.30	0.00
12	0.21	0.06	13.07	13.33		1	0.21	0.00
13	0.15	0.04	13.06	13.25		1	0.15	0.00
14	0.12	0.03	13.05	13.21		1	0.12	0.00
15	0.11	0.02	13.05	13.18		1	0.11	0.00
16	0.10	0.02	13.04	13.16		1	0.10	0.00
17	0.09	0.01	13.04	13.15		1	0.09	0.00
18	0.10	0.01	13.03	13.14		1	0.10	0.00
19	0.10	0.01	13.03	13.14		1	0.10	0.00
20	0.11	0.01	13.02	13.15		1	0.11	0.00
21	0.12	0.02	13.02	13.16		1	0.12	0.00
22	0.13	0.02	13.03	13.18		1	0.13	0.00
23	0.14	0.02	13.04	13.20		1	0.14	0.00
24	0.15	0.02	13.05	13.23		1	0.15	0.00
25	0.17	0.03	13.08	13.27		1	0.17	0.01
26	0.20	0.03	13.11	13.34		1	0.19	0.01
27	0.24	0.04	13.14	13.42		1	0.23	0.01
28	0.30	0.05	13.19	13.55		1	0.29	0.01
29	0.41	0.06	13.25	13.72		1	0.39	0.02
30	0.55	0.08	13.32	13.94		1	0.52	0.03
31	0.74	0.09	13.39	14.23		1	0.71	0.03

Table B.5 (Continued)

INSTANTANEOUS ENERGY (JOULES) OF R. SHANK WN21F2 NORMAL WALKING SPEED FILTERED AT 4 HZ TRUNK & FORCES MOD.

POS	TRANSLATIONAL	ROTATIONAL	POTENTIAL	TOTAL	MODE1	MODE2	X KINETIC	Y KINETIC
32	1.00	0.11	13.48	14.59		1	0.97	0.03
33	1.35	0.14	13.56	15.05		1	1.32	0.03
34	1.82	0.17	13.64	15.62		1	1.78	0.03
35	2.41	0.20	13.72	16.33		1	2.38	0.03
36	3.14	0.24	13.79	17.17		1	3.12	0.03
37	4.04	0.28	13.86	18.18	HSL	1	4.01	0.03
38	5.12	0.31	13.94	19.36		1	5.09	0.03
39	6.39	0.32	14.02	20.73		1	6.35	0.04
40	7.81	0.31	14.12	22.24		1	7.75	0.06
41	9.27	0.28	14.24	23.79		1	9.19	0.08
42	10.62	0.21	14.37	25.21		1	10.52	0.10
43	11.73	0.14	14.52	26.39		1	11.62	0.11
44	12.56	0.06	14.67	27.29	TOR	1	12.44	0.12
45	13.12	0.01	14.82	27.95		1	13.01	0.10
46	13.45	0.00	14.96	28.42		1	13.38	0.08
47	13.61	0.04	15.07	28.72		1	13.56	0.04
48	13.64	0.13	15.15	28.92		1	13.62	0.02
49	13.62	0.25	15.18	29.05		1	13.62	0.00
50	13.61	0.39	15.17	29.18		1	13.61	0.01
51	13.63	0.53	15.12	29.27		1	13.61	0.03
52	13.63	0.64	15.02	29.30		1	13.57	0.06
53	13.55	0.74	14.90	29.20		1	13.47	0.09
54	13.38	0.82	14.76	28.96		1	13.27	0.11
55	13.11	0.89	14.60	28.59		1	12.98	0.13
56	12.73	0.93	14.44	28.09		1	12.60	0.13
57	12.23	0.95	14.28	27.46		1	12.11	0.12
58	11.61	0.94	14.13	26.68		1	11.51	0.09
59	10.87	0.90	14.01	25.78		1	10.80	0.07
60	10.04	0.83	13.90	24.78		1	10.00	0.05
61	9.18	0.72	13.81	23.71		1	9.15	0.03
62	8.31	0.57	13.75	22.63		1	8.29	0.02
63	7.48	0.40	13.69	21.58		1	7.47	0.01
64	6.70	0.24	13.64	20.58		1	6.69	0.01
65	5.97	0.11	13.59	19.66		1	5.95	0.01
66	5.28	0.02	13.53	18.84		1	5.26	0.02
67	4.66	0.00	13.46	18.12		1	4.63	0.03
68	4.09	0.02	13.38	17.49		1	4.05	0.04
69	3.58	0.07	13.28	16.93	HSR	1	3.53	0.05
70	3.14	0.13	13.18	16.45		1	3.09	0.05

Table B.5 (Continued)

INSTANTANEOUS ENERGY (JOULES) OF R. SHANK   WN21F2   NORMAL WALKING SPEED   FILTERED AT 4 HZ   TRUNK & FORCES MOD.

POS	TRANSLATIONAL	ROTATIONAL	POTENTIAL	TOTAL	MODE1	MODE2	X KINETIC	Y KINETIC
71	2.77	0.17	13.09	16.03		1	2.73	0.04
72	2.44	0.20	13.01	15.64		1	2.41	0.02
73	2.12	0.21	12.95	15.28		1	2.11	0.01
74	1.81	0.21	12.92	14.93		1	1.80	0.00
75	1.49	0.19	12.89	14.58		1	1.49	0.00
76	1.18	0.17	12.88	14.23		1	1.18	0.00
77	0.90	0.14	12.87	13.91		1	0.90	0.00
78	0.67	0.11	12.85	13.64		1	0.67	0.00
79	0.50	0.09	12.84	13.43		1	0.50	0.00
80	0.38	0.07	12.82	13.27		1	0.38	0.00
81	0.29	0.05	12.81	13.15		1	0.29	0.00
82	0.23	0.04	12.80	13.06	TOL	1	0.22	0.00
83	0.17	0.03	12.79	12.99		1	0.17	0.00
84	0.13	0.02	12.79	12.94		1	0.13	0.00
85	0.10	0.02	12.79	12.91		1	0.10	0.00
86	0.08	0.01	12.80	12.89		1	0.08	0.00
87	0.07	0.01	12.80	12.89		1	0.07	0.00
88	0.07	0.01	12.81	12.89		1	0.07	0.00
89	0.07	0.01	12.83	12.90		1	0.07	0.00
90	0.07	0.01	12.84	12.92		1	0.07	0.00
91	0.08	0.01	12.87	12.95		1	0.07	0.00
92	0.09	0.01	12.89	13.00		1	0.09	0.00
93	0.11	0.02	12.92	13.06		1	0.11	0.01
94	0.14	0.02	12.96	13.13		1	0.13	0.01
95	0.18	0.03	13.01	13.23		1	0.17	0.01
96	0.24	0.04	13.07	13.34		1	0.22	0.02
97	0.32	0.05	13.13	13.49		1	0.30	0.02
98	0.44	0.06	13.19	13.69		1	0.42	0.02
99	0.62	0.08	13.26	13.96		1	0.59	0.02
100	0.86	0.10	13.33	14.30		1	0.84	0.02
101	1.19	0.12	13.40	14.72		1	1.17	0.02
102	1.60	0.15	13.47	15.21		1	1.57	0.02
103	2.04	0.17	13.53	15.74	HSL	1	2.02	0.02
104	2.45	0.19	13.59	16.23		1	2.43	0.02
105	2.71	0.19	13.65	16.55		1	2.69	0.02
106	2.69	0.18	13.71	16.58		1	2.67	0.02
107	2.33	0.14	13.77	16.24		1	2.31	0.02
108	1.72	0.10	13.82	15.64		1	1.71	0.01
109	1.13	0.06	13.86	15.05		1	1.13	0.01

Table B.6 Kinetic, Potential and Total Energy—Right Thigh

INSTANTANEOUS ENERGY (JOULES) OF R. THIGH   WN21F2   NORMAL WALKING SPEED   FILTERED AT 4 HZ   TRUNK & FORCES MOD.

POS	TRANSLATIONAL	ROTATIONAL	POTENTIAL	TOTAL	MODE1	MODE2	X KINETIC	Y KINETIC
1	9.42	0.00	58.92	68.33	HSR	1	9.40	0.02
2	11.47	0.00	58.86	70.33		1	11.46	0.00
3	12.74	0.00	58.87	71.61		1	12.74	0.00
4	13.05	0.00	58.94	71.99		1	13.02	0.02
5	12.43	0.00	59.07	71.51		1	12.38	0.05
6	11.14	0.01	59.24	70.39		1	11.07	0.07
7	9.52	0.02	59.42	68.96		1	9.44	0.07
8	7.90	0.05	59.60	67.54		1	7.83	0.07
9	6.49	0.09	59.76	66.34		1	6.43	0.06
10	5.38	0.14	59.92	65.44		1	5.32	0.06
11	4.55	0.19	60.08	64.82		1	4.49	0.06
12	3.94	0.23	60.23	64.40	TOL	1	3.89	0.05
13	3.48	0.25	60.38	64.11		1	3.43	0.05
14	3.13	0.26	60.52	63.90		1	3.09	0.04
15	2.84	0.25	60.64	63.73		1	2.81	0.03
16	2.61	0.24	60.74	63.60		1	2.59	0.02
17	2.45	0.23	60.82	63.50		1	2.44	0.01
18	2.35	0.22	60.86	63.43		1	2.35	0.00
19	2.33	0.20	60.87	63.40		1	2.33	0.00
20	2.36	0.19	60.86	63.40		1	2.36	0.00
21	2.41	0.18	60.82	63.42		1	2.41	0.00
22	2.49	0.17	60.78	63.44		1	2.48	0.00
23	2.58	0.17	60.74	63.49		1	2.57	0.01
24	2.70	0.17	60.69	63.56		1	2.70	0.01
25	2.88	0.17	60.63	63.69		1	2.87	0.01
26	3.11	0.17	60.58	63.86		1	3.11	0.01
27	3.41	0.16	60.52	64.09		1	3.40	0.01
28	3.77	0.15	60.46	64.38		1	3.77	0.01
29	4.21	0.13	60.40	64.74		1	4.20	0.01
30	4.73	0.10	60.34	65.17		1	4.72	0.01
31	5.36	0.07	60.27	65.70		1	5.34	0.01

Table B.6 (Continued)

INSTANTANEOUS ENERGY (JOULES) OF R. THIGH WN21F2 NORMAL WALKING SPEED FILTERED AT 4 HZ TRUNK & FORCES MOD.

POS	TRANSLATIONAL	ROTATIONAL	POTENTIAL	TOTAL	MODE1	MODE2	X KINETIC	Y KINETIC
32	6.13	0.05	60.19	66.37		1	6.11	0.02
33	7.09	0.02	60.09	67.21		1	7.06	0.03
34	8.26	0.01	59.96	68.22		1	8.20	0.05
35	9.60	0.00	59.79	69.40		1	9.52	0.08
36	11.08	0.01	59.58	70.67	HSL	1	10.96	0.12
37	12.63	0.03	59.34	71.99		1	12.47	0.16
38	14.15	0.07	59.07	73.29		1	13.98	0.18
39	15.58	0.14	58.78	74.51		1	15.41	0.18
40	16.80	0.23	58.52	75.55		1	16.66	0.15
41	17.67	0.33	58.29	76.29		1	17.57	0.10
42	18.07	0.43	58.11	76.61		1	18.03	0.04
43	18.01	0.51	58.02	76.53		1	18.00	0.00
44	17.62	0.55	58.02	76.19	TOR	1	17.61	0.00
45	17.08	0.55	58.14	75.77		1	17.01	0.07
46	16.46	0.52	58.37	75.36		1	16.28	0.18
47	15.80	0.47	58.70	74.97		1	15.48	0.32
48	15.08	0.41	59.11	74.59		1	14.65	0.44
49	14.31	0.34	59.56	74.21		1	13.81	0.50
50	13.52	0.27	60.03	73.82		1	13.02	0.44
51	12.71	0.22	60.49	73.41		1	12.26	0.44
52	11.87	0.17	60.90	72.94		1	11.51	0.35
53	11.00	0.13	61.26	72.39		1	10.75	0.25
54	10.13	0.09	61.55	71.78		1	9.99	0.14
55	9.32	0.06	61.76	71.14		1	9.26	0.06
56	8.59	0.04	61.88	70.52		1	8.58	0.02
57	7.98	0.02	61.93	69.93		1	7.98	0.01
58	7.49	0.01	61.90	69.40		1	7.48	0.01
59	7.15	0.00	61.80	68.95		1	7.10	0.04
60	6.95	0.00	61.63	68.58		1	6.86	0.09
61	6.95	0.00	61.40	68.35		1	6.80	0.15
62	7.16	0.00	61.12	68.28		1	6.96	0.19
63	7.58	0.00	60.82	68.40		1	7.37	0.21
64	8.18	0.00	60.52	68.69		1	7.97	0.20
65	8.87	0.00	60.23	69.11		1	8.70	0.17
66	9.59	0.00	59.97	69.56		1	9.45	0.14
67	10.23	0.00	59.74	69.97	HSR	1	10.12	0.11
68	10.71	0.00	59.53	70.24		1	10.63	0.08
69	10.99	0.00	59.36	70.34		1	10.93	0.05
70	11.06	0.00	59.23	70.29		1	11.04	0.03

Table B.6 (Continued)

INSTANTANEOUS ENERGY (JOULES) OF R. THIGH   WN21F2   NORMAL WALKING SPEED   FILTERED AT 4 HZ   TRUNK & FORCES MOD.

POS	TRANSLATIONAL	ROTATIONAL	POTENTIAL	TOTAL	MODE1	MODE2	X KINETIC	Y KINETIC
71	10.96	0.00	59.15	70.11		1	10.96	0.01
72	10.68	0.00	59.13	69.81		1	10.68	0.00
73	10.20	0.01	59.18	69.38		1	10.19	0.01
74	9.53	0.01	59.27	68.81		1	9.51	0.02
75	8.70	0.02	59.38	68.10		1	8.67	0.03
76	7.76	0.04	59.50	67.30		1	7.73	0.03
77	6.80	0.06	59.62	66.47		1	6.77	0.03
78	5.89	0.09	59.72	65.70		1	5.87	0.02
79	5.10	0.12	59.82	65.03		1	5.08	0.02
80	4.44	0.14	59.90	64.48	TOL	1	4.42	0.02
81	3.89	0.16	59.98	64.03		1	3.87	0.02
82	3.43	0.18	60.07	63.68		1	3.42	0.02
83	3.05	0.19	60.15	63.39		1	3.03	0.02
84	2.73	0.21	60.24	63.18		1	2.72	0.02
85	2.47	0.21	60.32	63.01		1	2.46	0.01
86	2.27	0.22	60.39	62.88		1	2.26	0.01
87	2.14	0.22	60.45	62.81		1	2.13	0.01
88	2.06	0.22	60.49	62.77		1	2.06	0.00
89	2.05	0.22	60.51	62.78		1	2.05	0.00
90	2.09	0.23	60.51	62.82		1	2.09	0.00
91	2.17	0.22	60.50	62.89		1	2.17	0.00
92	2.29	0.22	60.48	62.99		1	2.29	0.00
93	2.44	0.20	60.46	63.10		1	2.44	0.00
94	2.61	0.18	60.44	63.23		1	2.61	0.00
95	2.82	0.16	60.42	63.39		1	2.82	0.00
96	3.09	0.14	60.40	63.62		1	3.08	0.00
97	3.47	0.11	60.38	63.95		1	3.46	0.00
98	4.00	0.09	60.34	64.44		1	4.00	0.00
99	4.72	0.07	60.30	65.09		1	4.71	0.01
100	5.62	0.05	60.23	65.90		1	5.60	0.01
101	6.65	0.03	60.14	66.82		1	6.63	0.03
102	7.74	0.01	60.01	67.76		1	7.69	0.05
103	8.72	0.00	59.86	68.58	HSL	1	8.65	0.07
104	9.39	0.01	59.67	69.06		1	9.30	0.09
105	9.47	0.01	59.47	68.94		1	9.37	0.09
106	8.70	0.01	59.26	67.98		1	8.61	0.10
107	7.06	0.02	59.06	66.14		1	6.97	0.09
108	4.90	0.02	58.88	63.80		1	4.84	0.06
109	2.96	0.02	58.74	61.72		1	2.93	0.04

Table B.7 Kinetic, Potential and Total Energy—(H.A.T.)

INSTANTANEOUS ENERGY (JOULES) OF H-A-T   WN21F2 NORMAL WALKING SPEED   FILTERED AT 4 HZ   TRUNK & FORCES MOD.

POS	TRANSLATIONAL	ROTATIONAL	POTENTIAL	TOTAL	MODE1	MODE2	X KINETIC	Y KINETIC
1	56.55	0.14	660.16	716.84	HSR	1	56.34	0.21
2	70.38	0.14	659.42	729.95		1	70.23	0.16
3	79.83	0.13	656.80	736.77		1	79.74	0.09
4	83.57	0.11	658.38	742.06		1	83.54	0.02
5	82.03	0.07	658.28	740.39		1	82.03	0.01
6	77.05	0.04	658.64	735.72		1	76.92	0.13
7	70.90	0.01	659.50	730.41		1	70.47	0.43
8	65.39	0.00	660.87	726.26		1	64.53	0.87
9	61.47	0.00	662.67	724.16		1	60.12	1.35
10	59.29	0.09	664.82	724.21		1	57.49	1.80
11	58.54	0.22	667.24	726.00	TOL	1	56.37	2.17
12	58.69	0.41	669.84	728.94		1	56.25	2.43
13	59.16	0.69	672.55	732.39		1	56.59	2.57
14	59.42	1.01	675.30	735.72		1	56.88	2.53
15	59.06	1.33	677.97	738.36		1	56.81	2.25
16	58.00	1.60	680.41	740.00		1	56.27	1.73
17	56.42	1.75	682.44	740.61		1	55.35	1.07
18	54.60	1.78	683.92	740.30		1	54.12	0.48
19	52.80	1.69	684.79	739.28		1	52.68	0.12
20	51.18	1.51	685.10	737.80		1	51.18	0.00
21	49.75	1.30	684.94	735.99		1	49.70	0.04
22	48.50	1.09	684.38	733.97		1	48.32	0.18
23	47.58	0.88	683.50	731.97		1	47.23	0.35
24	47.30	0.71	682.36	730.37		1	46.77	0.53
25	47.77	0.58	681.03	729.38		1	47.09	0.68
26	48.89	0.49	679.55	728.93		1	48.08	0.80
27	50.42	0.42	677.98	728.81		1	49.50	0.91
28	52.18	0.37	676.30	728.86		1	51.16	1.02
29	53.95	0.32	674.54	728.81		1	52.83	1.12
30	55.44	0.26	672.70	728.40		1	54.25	1.19
31	56.63	0.18	670.83	727.64		1	55.42	1.21

Table B.7 (Continued)

INSTANTANEOUS ENERGY (JOULES) OF H-A-T   WN21F2   NORMAL WALKING SPEED   FILTERED AT 4 HZ   TRUNK & FORCES MOD.

POS	TRANSLATIONAL	ROTATIONAL	POTENTIAL	TOTAL	MODE1	MODE2	X KINETIC	Y KINETIC
32	57.90	0.10	668.96	726.95		1	56.72	1.18
33	59.69	0.04	667.13	726.87		1	58.60	1.09
34	62.20	0.01	665.40	727.61		1	61.24	0.96
35	65.13	0.00	663.79	728.92		1	64.33	0.80
36	67.70	0.07	662.35	730.07		1	67.08	0.62
37	68.97	0.07	661.11	730.14		1	68.54	0.42
38	68.33	0.21	660.14	728.68	HSL	1	68.11	0.22
39	65.94	0.45	659.50	725.89		1	65.87	0.07
40	62.37	0.77	659.22	722.35		1	62.36	0.00
41	58.18	1.08	659.30	718.57		1	58.16	0.02
42	53.76	1.34	659.75	714.84		1	53.63	0.13
43	49.54	1.50	660.54	711.59		1	49.21	0.33
44	46.16	1.56	661.70	709.42		1	45.53	0.62
45	44.02	1.52	663.23	708.77	FOR	1	43.01	1.02
46	43.09	1.39	665.13	709.61		1	41.61	1.48
47	43.01	1.24	667.37	711.61		1	41.06	1.95
48	43.35	1.07	669.88	714.30		1	41.03	2.32
49	43.73	0.91	672.55	717.20		1	41.25	2.48
50	44.00	0.76	675.24	720.00		1	41.66	2.34
51	44.19	0.60	677.76	722.56		1	42.26	1.94
52	44.43	0.43	679.98	724.83		1	43.05	1.38
53	44.82	0.27	681.76	726.84		1	44.00	0.82
54	45.47	0.13	683.05	728.66		1	45.09	0.38
55	46.44	0.04	683.86	730.34		1	46.31	0.13
56	47.55	0.00	684.25	731.81		1	47.53	0.02
57	48.59	0.01	684.30	732.90		1	48.58	0.00
58	49.42	0.04	684.03	733.50		1	49.36	0.06
59	50.07	0.08	683.44	733.59		1	49.86	0.21
60	50.56	0.09	682.49	733.14		1	50.11	0.45
61	50.94	0.07	681.16	732.16		1	50.15	0.79
62	51.35	0.02	679.46	730.84		1	50.19	1.16
63	51.97	0.00	677.49	729.46		1	50.52	1.45
64	52.84	0.04	675.37	728.25		1	51.30	1.54
65	53.94	0.15	673.26	727.34		1	52.50	1.44
66	55.36	0.26	671.28	726.90		1	54.14	1.22
67	57.27	0.34	669.49	727.10		1	56.29	0.97
68	59.65	0.35	667.92	727.92	HSR	1	58.89	0.76
69	62.34	0.32	666.53	729.18		1	61.74	0.59
70	65.01	0.25	665.29	730.56		1	64.56	0.45

Table B.7 (*Continued*)

INSTANTANEOUS ENERGY (JOULES) OF H-A-T   WN21F2   NORMAL WALKING SPEED   FILTERED AT 4 HZ   TRUNK & FORCES MOD.

POS	TRANSLATIONAL	ROTATIONAL	POTENTIAL	TOTAL	MODE1	MODE2	X KINETIC	Y KINETIC
71	67.21	0.19	664.23	731.63		1	66.90	0.31
72	68.36	0.13	663.41	731.89		1	68.20	0.15
73	68.20	0.08	662.90	731.17		1	68.16	0.03
74	66.95	0.03	662.78	729.76		1	66.95	0.00
75	65.11	0.01	663.07	728.18		1	65.02	0.09
76	63.14	0.00	663.79	726.93		1	62.85	0.30
77	61.50	0.03	664.92	726.45		1	60.90	0.60
78	60.55	0.12	666.44	727.11		1	59.59	0.97
79	60.26	0.31	668.27	728.83		1	58.95	1.30
80	60.03	0.56	670.32	730.91	IOL	1	58.48	1.54
81	59.15	0.79	672.49	732.44		1	57.51	1.64
82	57.40	0.94	674.68	733.03		1	55.78	1.62
83	55.08	1.01	676.82	732.91		1	53.58	1.50
84	52.63	1.03	678.86	732.52		1	51.34	1.29
85	50.36	1.05	680.69	732.11		1	49.38	0.98
86	48.49	1.08	682.23	731.80		1	47.87	0.62
87	47.21	1.11	683.37	731.70		1	46.92	0.29
88	46.59	1.13	684.07	731.80		1	46.52	0.07
89	46.54	1.13	684.30	731.97		1	46.54	0.00
90	46.83	1.10	684.08	732.01		1	46.77	0.06
91	47.20	1.03	683.46	731.69		1	46.97	0.22
92	47.37	0.94	682.47	730.78		1	46.93	0.44
93	47.19	0.83	681.19	729.22		1	46.52	0.67
94	46.73	0.72	679.68	727.13		1	45.86	0.87
95	46.23	0.61	678.01	724.85		1	45.23	1.01
96	46.00	0.48	676.26	722.74		1	44.92	1.08
97	46.31	0.35	674.47	721.13		1	45.21	1.09
98	47.34	0.22	672.70	720.25		1	46.29	1.05
99	49.14	0.11	670.99	720.24		1	48.18	0.97
100	51.66	0.03	669.35	721.04		1	50.77	0.89
101	54.60	0.00	667.78	722.38		1	53.78	0.82
102	57.37	0.00	666.26	723.63		1	56.60	0.77
103	59.08	0.02	664.80	723.89		1	58.38	0.70
104	58.66	0.04	663.41	722.10	HSL	1	58.04	0.61
105	54.94	0.07	662.13	717.13		1	54.44	0.50
106	47.13	0.10	660.99	708.22		1	46.75	0.38
107	35.77	0.11	660.03	695.91		1	35.52	0.25
108	23.25	0.10	659.30	682.65		1	23.13	0.13
109	13.06	0.08	658.82	671.96		1	13.01	0.04

Table B.8   Total Body Energy, Integrated Absolute Energy Change

TOTAL ENERGIES (JOULES) OF ALL SEGMENTS

POS	TIME	TOTAL ENERGY	CUM.ABS.CHANGE	MODE1	MODE2
1	0.00000	892.17	0.00	HSR	1
2	0.01667	910.01	17.84		1
3	0.03334	922.77	30.60		1
4	0.05001	928.94	36.77		1
5	0.06668	929.08	36.91		1
6	0.08335	925.28	40.70		1
7	0.10002	920.15	45.83		1
8	0.11669	915.81	50.17		1
9	0.13336	913.52	52.47		1
10	0.15003	913.61	52.56	TOL	1
11	0.16670	915.70	54.66		1
12	0.18337	919.08	58.03		1
13	0.20004	922.97	61.92		1
14	0.21671	926.66	65.62		1
15	0.23338	929.58	68.53		1
16	0.25005	931.33	70.28		1
17	0.26672	931.78	70.74		1
18	0.28339	930.98	71.54		1
19	0.30006	929.15	73.37		1
20	0.31673	926.59	75.93		1
21	0.33340	923.48	79.04		1
22	0.35007	919.95	82.58		1
23	0.36674	916.24	86.28		1
24	0.38341	912.78	89.74		1
25	0.40008	909.86	92.67		1
26	0.41675	907.49	95.03		1
27	0.43342	905.59	96.94		1
28	0.45009	904.07	98.45		1
29	0.46676	902.77	99.75		1
30	0.48343	901.48	101.05		1
31	0.50010	900.28	102.25		1
32	0.51677	899.62	102.91		1
33	0.53344	900.01	103.29		1
34	0.55011	901.61	104.90		1
35	0.56678	904.12	107.41		1
36	0.58345	906.72	110.01	HSL	1
37	0.60012	908.41	111.70		1
38	0.61679	908.66	111.95		1
39	0.63346	907.61	112.99		1
40	0.65013	905.77	114.84		1
41	0.66680	903.46	117.14		1
42	0.68347	900.87	119.74		1
43	0.70014	898.39	122.22		1
44	0.71681	896.80	123.80		1
45	0.73348	896.72	123.89	TOR	1
46	0.75015	898.14	125.31		1
47	0.76682	900.68	127.85		1
48	0.78349	903.82	130.98		1
49	0.80016	907.07	134.24		1
50	0.81683	910.14	137.30		1
51	0.83350	912.82	139.98		1
52	0.85017	914.97	142.14		1
53	0.86684	916.53	143.69		1
54	0.88351	917.58	144.75		1
55	0.90018	918.25	145.42		1
56	0.91685	918.50	145.66		1
57	0.93352	918.17	145.98		1
58	0.95019	917.18	146.98		1
59	0.96686	915.51	148.65		1
60	0.98353	913.21	150.94		1
61	1.00020	910.43	153.73		1
62	1.01687	907.50	156.66		1
63	1.03354	904.84	159.31		1
64	1.05021	902.77	161.39		1
65	1.06688	901.48	162.68		1
66	1.08355	901.13	163.03		1
67	1.10021	901.78	163.67		1
68	1.11688	903.14	165.03		1
69	1.13355	904.65	166.54	HSR	1
70	1.15022	905.54	167.44		1
71	1.16689	904.94	168.04		1
72	1.18356	902.20	170.78		1
73	1.20023	897.58	175.40		1
74	1.21690	892.40	180.58		1

Index

DATE DUE